W0225618

Logistic Support of a Manned
Underwater Production Complex

Logistic Support of a Manned Underwater Production Complex

MICHAEL E. JONES

Graham & Trotman

First published in 1983 by
Graham & Trotman Ltd.
Sterling House,
66 Wilton Road,
London SW1V 1DE

British Library Cataloguing in Publication Data

Jones, Michael E.W.
Logistic support of a manned underwater production
complex.
1. Oil well drilling, Submarine
I. Title
622'.3382   TN871.3

ISBN 0-86010-450-8

Softcover reprint of the hardcover 1st edition 1983

ISBN-13: 978-94-009-6657-4   e-ISBN-13: 978-94-009-6655-0

DOI: 10.1007/ 978-94-009-6655-0

by Billings & Sons Limited, Worcester.
Typeset in Great Britain by Academic Typing Service,
Gerrards Cross, Bucks.

CONTENTS

LIST OF FIGURES

LIST OF TABLES

ACKNOWLEDGEMENTS

I should like to acknowledge the financial support provided for this work by the Marine Technology Directorate of the Science and Engineering Research Council as part of the Marine Technology Programme at the University of Glasgow.

I also acknowledge the permission provided by McGraw-Hill and Wiley & Sons, publishing companies, to include drawings from references 2 and 3, respectively.

Finally I wish to express my appreciation to Isobella Lawson for the preparation of the drawings for the book and for typing the final manuscript.

1
INTRODUCTION

The support of subsea oil and gas production operations involves the use of many underwater work systems. Divers can be used for support tasks in water depths to 300 m, but at more extreme depths operations become restrictively expensive and the efficiency of task performance is reduced. Remote controlled unmanned vehicles can replace the diver to a limited extent, performing inspection and maintenance tasks and supporting drilling operations. Operations in deepwaters performed by remote controlled vehicles and one man submersible vehicles, such as JIM and WASP, are more cost effective than the use of divers. The areas of operation of the more complex multi-manned submersibles and bells are today generally restricted to their use for diver lock-out operations, manned intervention to subsea enclosures and the deployment of other underwater work systems.

Oil and gas exploration activity is being undertaken in progressively deeper waters. In the North Sea, Shell have discovered a large gas accumulation off the Norwegian coast in 323 m water depth and B.P. have made oil finds West of the Shetlands in 500 m and West of Eire in 450 m. Exploration drilling is today being carried out in many areas of the world in water depths greater than 1000 m, i.e. Western Mediterranean, Offshore Argentina, Offshore Western Australia and in the Niger Basin, West Africa. The existing discoveries of Shell and B.P., and any exploitable

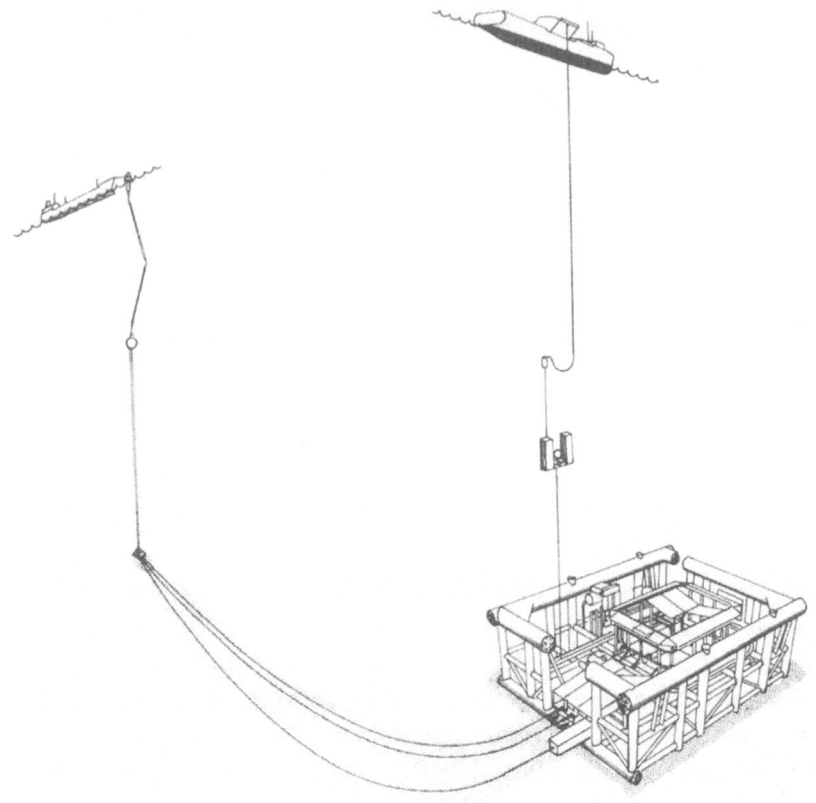

Figure 1.1 Exxon submerged production system

finds in deeper waters, will require new production technology to develop such fields.

In such deepwaters, where surface piercing systems experience mooring and riser problems, and floating production systems exhibit payload limitations in the development of larger reservoirs, the only options available for deepwater production may be the use of complex subsea production systems. On the basis of contemporary developments in subsea production systems, such systems are likely to be available in two configurations. One, the remote controlled 'wet' ambient pressure subsea production complex and the other, the manned one atmosphere encapsulated subsea production system.

In the offshore oil industry at the present time, there appears to be a firm trend towards 'wet' ambient pressure subsea completions, and progressive developments in this area have led to the engineering design and operation of the complex remotely operated and maintained 'wet' manifold and production systems, epitomised in the Exxon Submerged Production System (Fig. 1.1) and the Elf Aquitaine subsea production system on the Grondin Field, off West Africa. Both systems use highly sophisticated technology for remote control of the subsea production operations and for the deployment of remote manipulators from the surface for repair and maintenance operations.

The development of the one atmosphere completion systems and manifold centres has proceeded in parallel with that of the 'wet' equivalent systems. The acceptance of such systems by the industry has been slow, probably due to the excessive initial costs of such systems and the relatively shallow water depths of present offshore production operations.

The major advantages of the one atmosphere systems are that relatively conventional surface equipment can be enclosed in a structure and placed in the underwater environment, and that manned intervention into the subsea systems is possible at one atmosphere, using underwater vehicles, i.e. submersibles or bells. The initial cost of such a system may be more expensive; however, maintenance and repair operations are easier and experienced oil-field technicians can be deployed to the wellhead or manifold with the appropriate tool packages. In deepwaters, where diver deployment is not physiologically possible or not cost effective, and where the limited work capabilities of submersible vehicles are restrictive, the one atmosphere system comes into its own.

In the seventies, a complex of one atmosphere subsea wells and a manifold centre, were installed on the Garoupa Field, offshore Brazil, for Petrobras by CanOcean Resources (Fig. 1.2). From such beginnings, the concept of the total subsea production system has evolved. This production system does not entail the development of new sophisticated production technology as the remote controlled 'wet' systems require, but rather relocates surface hardware, whose maintenance and reliability characteristics are well proven, in a similar environment subsea. The facility for manned intervention into that environment provides for *in-situ* monitoring, repair and maintenance of the operating equipment and reduces dependence on surface systems for operation. Manned interven-

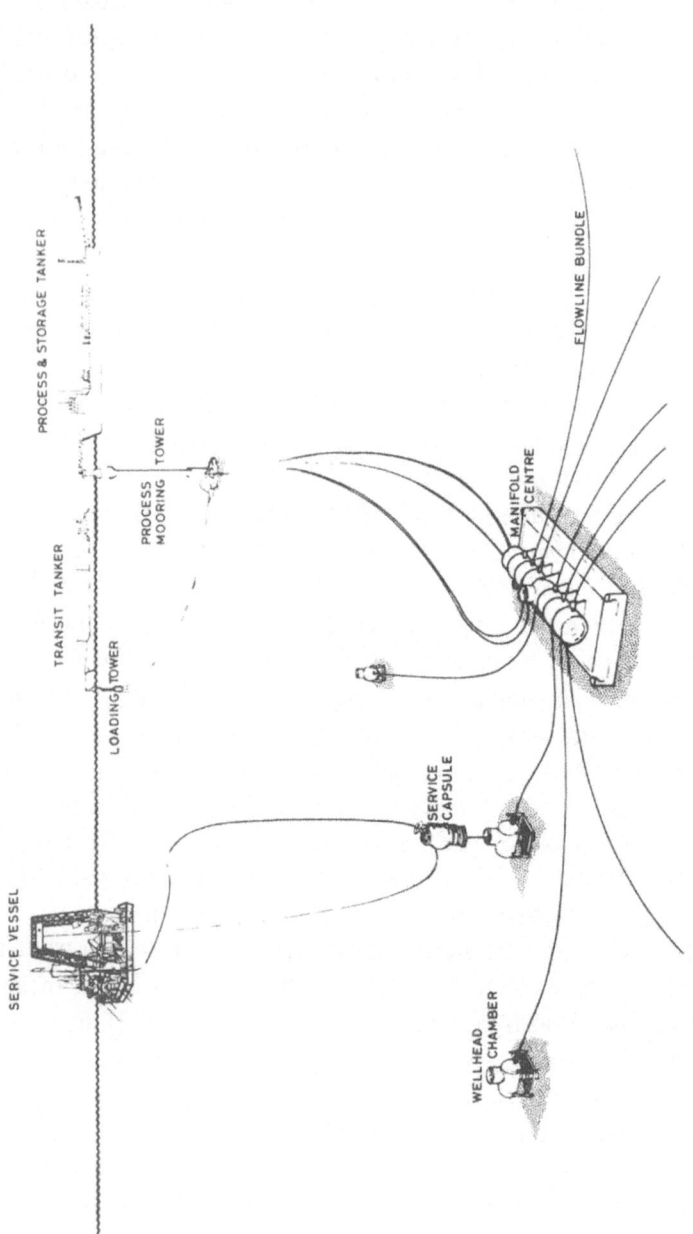

Figure 1.2 Garoupa Field development — CanOcean Resources

tion allows the crew to bring a level of adaptability to the task requirements, which provides the best insurance for reliability and maintainability.

The fundamental problem involved in the selection of either the 'wet' or 'dry' subsea production system, is based in the need to install, maintain and repair the operating system on the seabed over the many years of the field life. The oil companies generally prefer direct manned access to all operations, including monitoring and maintenance and, if at all possible, in the 'dry' rather than the 'wet'. The development of the 'wet' total subsea production system is well in advance of the one atmosphere production system, with detailed engineering trials already undertaken. The operation of remote control equipment, however, requires the incorporation of sophisticated technology, which raises reliability problems. System design can take account of some failures and incorporate maintenance systems to correct predicted malfunctions, which further complicates the total system engineering.

The operating experience of such automatic subsea production systems in the near future (Shell have installed a complex underwater manifold, based on Exxon's experience with the prototype submerged production system on the South Cormorant field) will determine the reliability of such engineering designs and whether they will provide the continuous production flow so critical to the industry. This experience from shallow water operations will indicate the future for the alternative concept, the one atmosphere production system in deep waters.

Manned underwater structures may be used for the subsea production of oil and gas in deep waters (Fig. 1.3) or for other oceanographic tasks in water depths beyond the limitations of hyperbaric diving activities. The deployment to, and the support of such structures on the seabed will be a major consideration in the effective and safe operation of such a facility. The use of one atmosphere subsea completion systems and underwater manifold centres with intermittent intervention into these systems, by the use of manned submersibles and bells, is at present established practice in the offshore oil industry. This book therefore, addresses itself to a detailed investigation of the problems of providing logistic support to a manned large one atmosphere subsea production complex.

A specific design concept has not been finalised and the exact

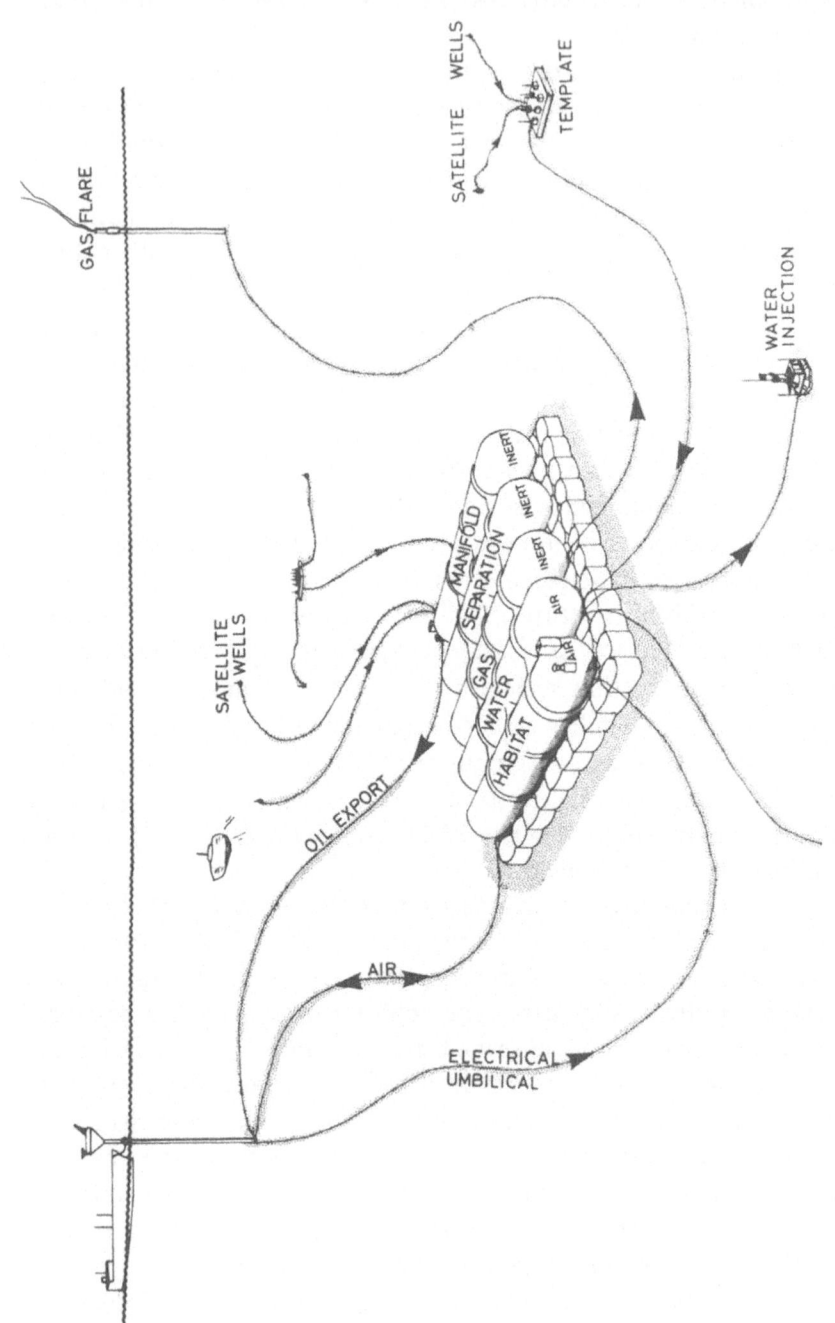

Figure 1.3 Deep sea production complex. (Also on opposite page.)

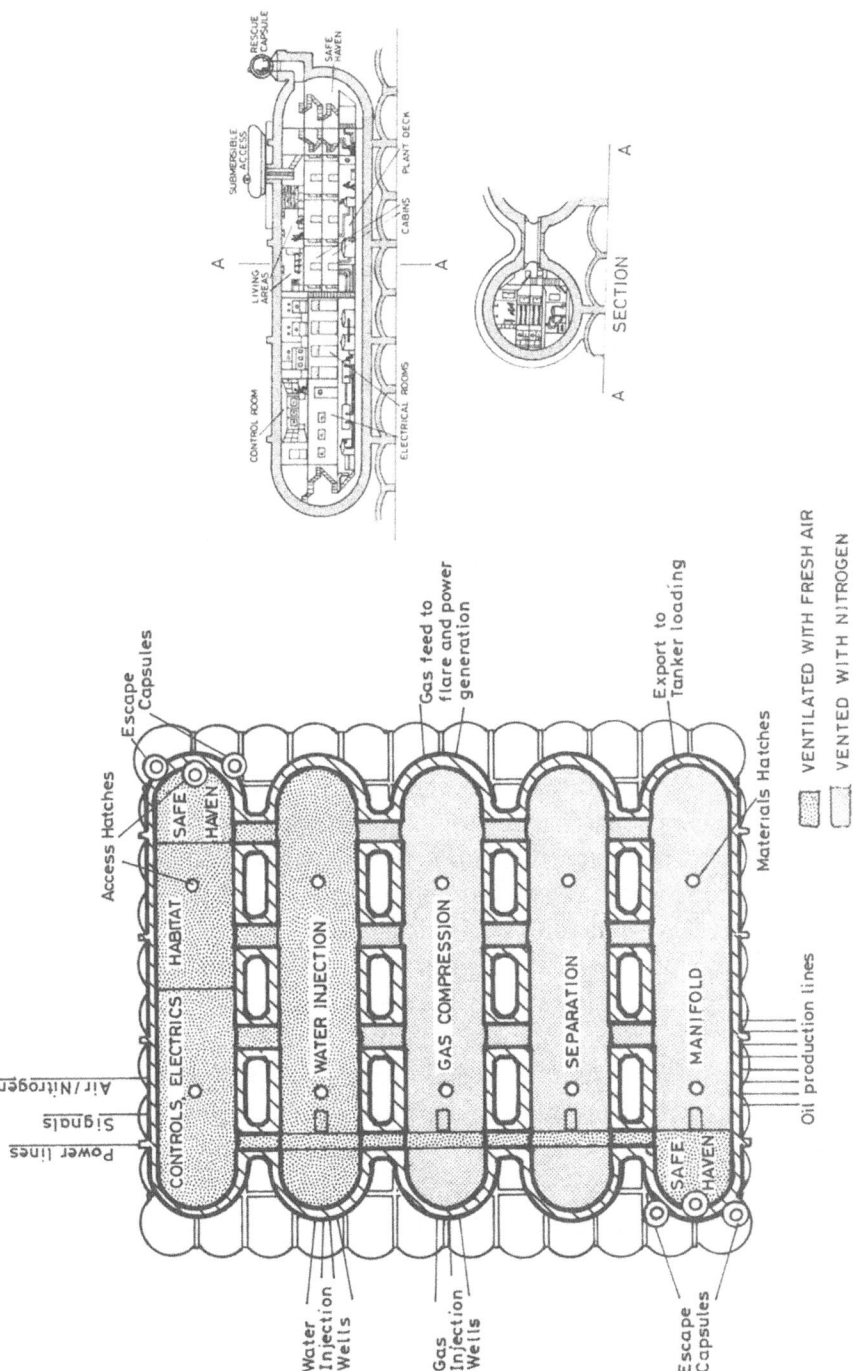

site location and field parameters are undetermined for the use of such a production concept, therefore a generalised treatment of the logistic support requirements is used. Initially, the expected environment of operations is defined in detail and operational parameters identified.

Main and auxiliary task requirements that are required to support the underwater complex are constructed and system constraints to the operation of the various subsystems identified. Environmental constraints at the various interfaces are considered along with inherent subsystem limitations and production system restrictions. Candidate support systems to match the various operational requirements are assessed and critical operational areas classified for further detailed investigation.

As it is envisaged that the large underwater complex will be constructed in a dry dock and the total system floated out from the base and towed on the surface to the proposed offshore site, before being deployed from the surface through the water column to the seabed, the logistics involved in these operations are investigated.

The logistic payloads in terms of men and materials, that are to be transferred through the water column in support of the operation of the complex are evaluated in detail. The influence of the complex design parameters on the requirement for support operations and the associated costs of these operations is evaluated. A logistic support operating philosophy is adopted for the support operations through the water column to allow definition of the submersible vehicle required to perform the role.

Access to the underwater complex from the submersible support vehicle is investigated. Designs of suitable ingress/egress systems are proposed that will retain the integrity of the structure during transfer operations of crew and materials through the interface. Various systems are investigated that could improve operational safety and provide a secondary means of escape from the complex in an emergency.

Surface and subsea support systems are then considered in terms of production operations and the need for an underwater vehicle launch and recovery vessel. Launch and recovery techniques are evaluated in relation to expected environmental conditions and the probable weight and size of the submersible vehicle.

Conclusions are drawn from the areas investigated and recommendations for further development work made.

2

THE ENVIRONMENT
OF OPERATIONS

2.1 INTRODUCTION

The logistic support of a manned underwater structure requires the assimilation of a three-dimensional concept of the ocean environment. Environmental constraints on support operations will be imposed at the surface/air interface, through the water column and at the seafloor/water interface. The water depths of this application are roughly defined as from 200 m (from approximately the continental shelf break) down the continental slope and to depths approaching 2000 m. It is within this depth range that our investigations of the oceanic environment will be concentrated.

The ocean environment, which covers 70-71% of the earth's surface is a huge dynamic system which influences weather, climate, energy supplies, food supplies and other areas critical to mankind. The surface and subsea environmental conditions at a site location will be unique and a function of the geographical situation. Local environmental conditions will need to be established in detail and correlated with broader related factors such as tides and seasons.

The prevailing environmental conditions will have a direct influence on the design of the subsea structure and foundation system and the design and implementation of docking and mating

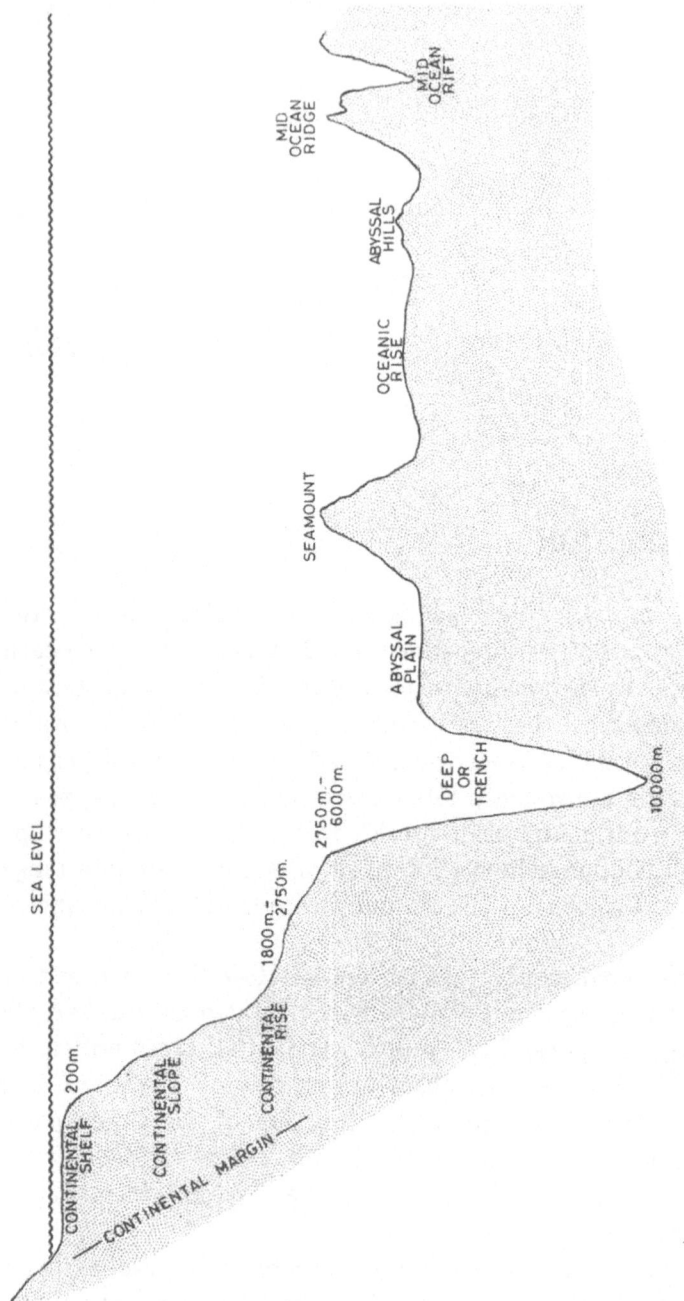

Figure 2.1 Geological features of the ocean floor (Ref. 3)

techniques for subsea support craft. Surface conditions will affect the successful operation of support vessels during the launch and recovery of submersible vehicles, tanker loading systems and any other surface based support equipment.

As no specific site has at present been chosen for the installation of such a subsea production complex, within the expected operating depth range and operational mode of such an underwater structure a general assessment of the prevailing environmental parameters is attempted.

2.2 THE OCEAN ENVIRONMENT

The ocean bottom topography has an even greater variety than the land surface, with deep canyons, high plateaux, volcanoes and long mountain ranges. The highest peaks of some undersea mountains are higher than Everest. A general profile of the ocean bottom is shown in Fig. 2.1. The oceans have an average depth of 4 km and, with horizontal distances of over 10,000 km, physically represent a thin film of water over the ocean basins.

In the subsea production of hydrocarbons we are generally concerned with operations on the continental margin. This is an extension of the land continent which protrudes under the sea and is composed of three major water depth regions. The continental shelf, at present a major area of oil exploitation in the North Sea, forms a shallow terrace of average slope 0.1° in water depths from 10 m to 300 m. The average depth of the continental shelves is less than 200 m and they cover approximately 8% of the ocean floor. At the continental shelf break, the continental slope falls away rapidly into deep water with an average slope of 4°. The continental slope accounts for 12-15% of the world's ocean bottom. At depths of the order of 2000 - 3000 m the continental slope levels off to an average slope of less than 0.1° and is normally termed the continental rise. The continental margins also contain many other irregularities which can include submarine canyons, basins and escarpments. The deep ocean basins, which only account for 10% of the ocean floor are normally located at the foot of the continental margin where they descended to depths of 10,000 m and form the point of origin of many subsea

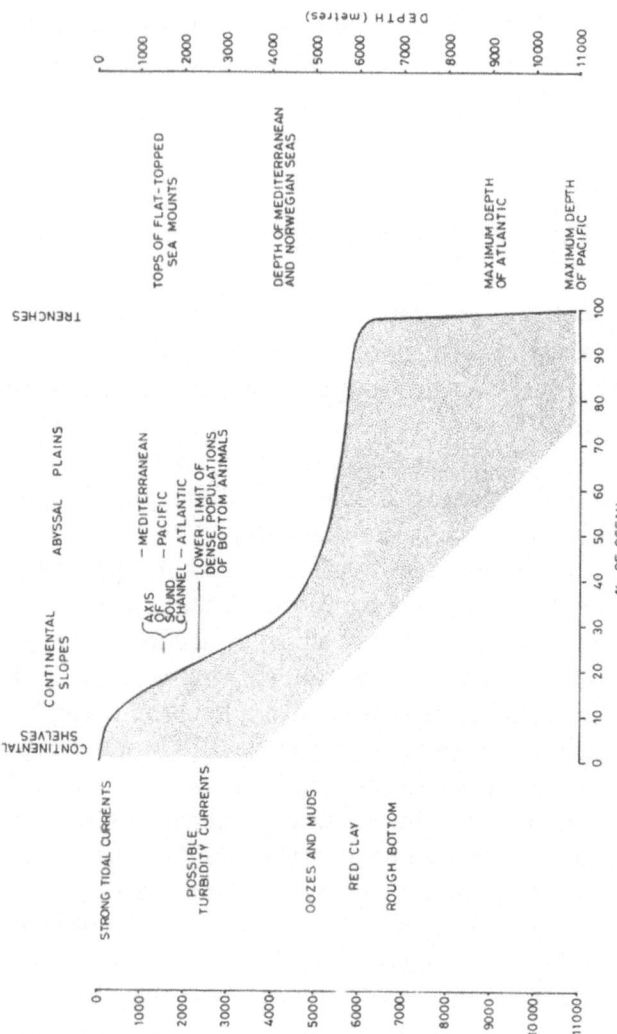

Figure 2.2 Hypsographic diagram — ocean depths (Ref. 2)

earthquakes. Beyond the margin lie the abyssal plains, oceanic rise and mid-ocean ridges of the ocean basin floor.

If we consider a depth range of application for structures to be 200–2000 m, the areas of application would cover most of the continental slope, shallow areas of the mid-ocean ridges, the tops of sea-mounts and the borders of atolls and sea-mounts. Table 2.1 shows a percentage breakdown of depth zones in the ocean (Svendug 1942) with adjacent seas included. Figure 2.2 gives a hypsographic representation of depth distribution and major features.

TABLE 2.1 % Areas of depth zones in the oceans

Depth interval (m)	Including adjacent seas			
	Atlantic	Pacific	Indian	All Oceans
0–200	13.3	5.7	4.2	7.6
200–1000	7.1	3.1	3.1	4.3
1000–2000	5.3	3.9	3.4	4.2
TOTAL	25.7	12.7	10.7	16.1

Interpretating from the table, it is apparent that approximately 8.5% of the world's oceans lie in the depth range of application, with 12.4% of the Atlantic, 7.0% of the Pacific and 6.5% of the Indian Ocean within the range. It has been established[1] that fixed offshore oil production systems may extend their depth of operations to the edge of the continental shelves at 300 m. The major areas of operation will therefore be based on the continental slope. These continental slopes have an average slope of 4° varying from 1° off North West Australia to 20° in the Gulf of California, and in some cases up to 45° locally. Both factors will restrict the area of application of seabed structures; however, they may also be installed on continental shelf areas in shallower depths where ice cover is restrictive (Arctic Ocean) or below busy shipping lanes.

The continental slopes are the major area of application, and in the past they have been neglected as sites for marine installations. The slopes are generally within reasonable distance of shore, however, they are frequently in areas of high current and heavy weather because of their exposed position. The slopes and associated rises are not adequately charted, except where close to areas of marine activity. Many are rugged and cut by large canyons and valleys. Mid-ocean ridges can project above the 300 m contour and are of mountainous topography with rock outcrops; oceanic islands and atolls are usually surrounded by very steep slopes, and as such, are probably not of interest for oil exploitation.

2.3 OCEAN ENVIRONMENTAL PARAMETERS: CONTINENTAL SLOPE

The installation and support of a manned underwater structure in the depth range 300–2000 m on the continental slope, will require a detailed assessment of the environmental factors that will affect operations at the site location. Parameters will vary as a function of the geographical location, season of year, month and even day. Oceanographic environmental parameters are investigated and their impact on operations indicated.

2.3.1 Pressure

The hydrostatic pressure exerted on the structure and over the depth range will vary with depth, but will be relatively invariant throughout the world. Pressure experienced will range from sea level, 1 atm (101.4kPA, 14.7psi) to 200 atm (20270kPA, 2940psi) at 2000 m. The pressure/depth gradient will vary slightly by up to 5% depending on the latitude, temperature and salinity. Pressure is of concern in the design of hulls for structures and submersible vehicles, hatches, mating systems, penetrations and a multitude of associated submarine systems.

2.3.2 Temperature

The temperature variation in the ocean is relatively small and can extend from about −1.8°C to 30°C. The highest temperatures are

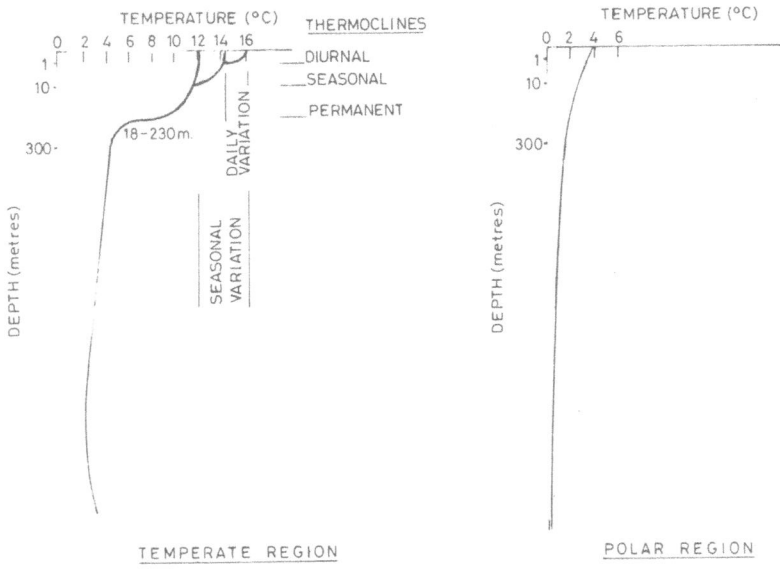

Figure 2.3 Graph of temperature versus depth (Ref. 3)

recorded at the sea surface and vary with the hour of the day. Daily and seasonal variations are most pronounced during the late summer months but are not easily distinguished in the late winter months. Figures 2.3 and 2.4 show the variation in various thermoclines, i.e. temperature gradients, in relation to day and seasonal changes and the permanent thermocline at deeper depths. The permanent thermocline exists below the zone where homogeneous mixing is continually going on down to about 100 m, beyond which there is a very rapid temperature drop with depth. In tropical climates the thermocline begins at shallower depths (50 m). Below the permanent thermocline, temperatures to the bottom fall very slowly and can be considered isothermal at close to freezing (2°C). Water temperatures in polar regions show little variation (range 0-4°C) and then only on the surface in summer months.

In the depth range considered, we may be operating in the permanent thermocline from 300 m down to about 1000 m and from there to 2000 m in an isothermal area close to freezing. Temperature variations will alter sea water density and salinity, causing changes in buoyancy and confusing underwater acoustic location and communications. Environmental conditioning equipment of the structure and submersibles will need to cope with the

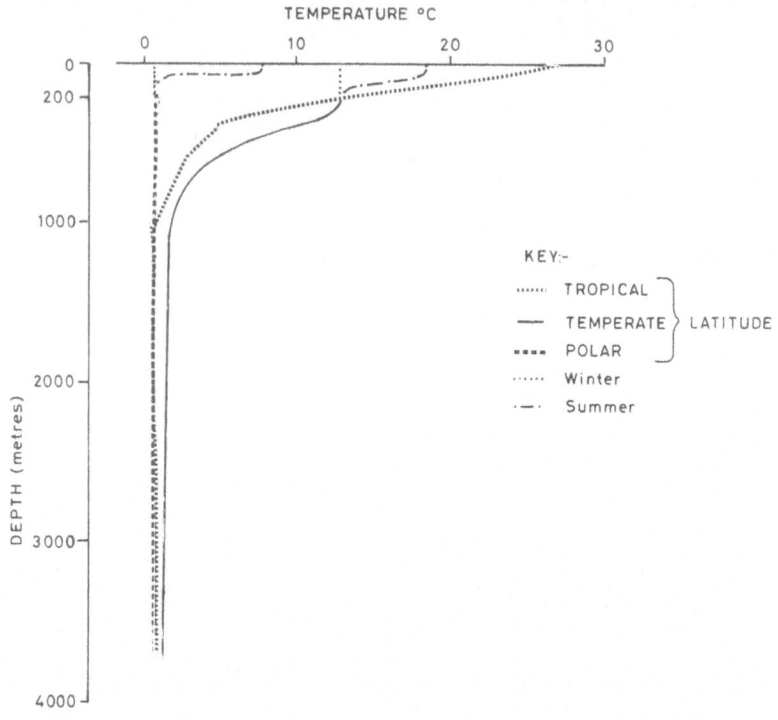

Figure 2.4 Graph of temperature versus depth (Ref. 2)

operating temperature range. This will be especially critical under emergency conditions, when crew heating for survival before rescue, is required.

2.3.3 Salinity

The salinity of the ocean averages out at about 35°/oo (3.5%) and may vary in range between 33 and 37° /oo. Salinity gradients will occur in thermoclines and will vary quite widely with geographical locations. Higher salinities will be recorded in high temperature areas such as the Red Sea, and lower values will be found where there are strong inflows of fresh water. Generally in temperate climates near-surface waters are usually warm and saline, while deeper waters are cool and less saline. Deep, less-saline water, when heated, will rise and remain at the higher level because it is less dense. This may have implications for cooling of the underwater structure's environment. Salinity may affect corrosion rates, but its effect on the density of seawater will be less than temperature.

2.3.4 Density

The density of seawater will vary with temperature, salinity, pressure and other factors such as mixing effects and evaporation. Density increases rapidly with depth, until, below the layer of rapid change, the density increases slowly due to the compression of sea water (Fig. 2.5). Temperature is a more important influence on density than salinity and it is dangerous to equate variations in salinity with density. The density of sea water from sea level to 2000 m will have a range of the order of 1031 up to 1036 kg/m³. Density will be critical to all subsea systems relying on buoyancy for their operation, and variations may disturb acoustic location and communications systems.

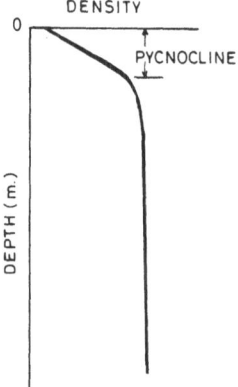

Figure 2.5 Change in density with depth (Ref. 3)

2.3.5 Currents

Current conditions will vary with geographical location but may be a combination of tidal, permanent and wind driven currents. Tidal currents caused by planetary motion can be complex and rotary in nature and may have significant values as flow passes from the deep oceanic region up the slope and across the shelf; speeds can vary between 0.25 and 2 knots. The topography of the slope may transform uniform general currents into erratic local eddies. Generally mid-water and deep water currents will not exceed one knot and in very deep waters will be even less, seafloor ripples will show where stronger currents have flowed in the past. Areas that concentrate flows such as the Straits of Gibraltar and Florida should be avoided, as higher current values will exist with the added complication of sheer currents setting up internal waves. The major permanent ocean currents such as the Kuroshia and Gulf Streams can reach speeds between 2 and 5 knots and are significantly affected by bottom topography.

Surface and near surface currents are quite well understood. However areas of convergence of winds, currents and temperature can cause waters to be forced together and sink, creating vertical currents and providing oxygen rich horizontal currents on reaching the seafloor. Conditions also combine to produce an up-welling cold current from the ocean depths, high in nutrients, like that off the coast of Peru.

Current conditions are important environmental parameters for subsea system design because the physical effects of current introduce hydrodynamic drag and produce drift of floating equipment and are a source of vertical and horizontal sheer force on vehicles and sub-surface operations. Surface currents will affect the drift of surface support vessels, buoys, and interfere with emplacement techniques. Sub-surface and seabed currents will affect submersible operations, emplacement, umbilicals, risers, control cables, seabed flowlines, visibility and the underwater structural design.

2.3.6 Turbidity Currents

Near seafloor turbidity currents, are not ocean currents in the true sense of the word, but more sediment laden streams or currents of mud, sand and stone. These heavy fluid masses react to gravity and behave like a submarine landslide flowing like an avalanche, the mass cascades down the continental slopes to the ocean floor. The rate of flow can be very rapid, speeds of 50 mph have been measured, and they can cover distances of up to 300 miles.

The currents are not constant and only occur when a build-up of sand, silt and mud sediment reaches a size when a flow is triggered. Large slides are probably less frequent than big earthquakes, however, local disturbances may be as common as small earthquakes, particularly in submarine canyons and channels. The catastrophic effect of such a phenomena on a submersible or even a large underwater structure in its path is clear. The siting of an underwater structure and the route for submersible support operations should avoid submarine canyons, crevices and ridges that offer potential for such a disaster.

2.3.7 Bottom Sediments

The continental slope being part of the continental margin and continental mass, will exhibit numerous rock outcroppings. Bottom sediments to 2000 m are not as varied as would be expected, as tidal scrubbing tends to remove finer sediments and leaves a relatively solid bottom. Organic materials are scarcer and anthigenic materials more common on the slope than the shelves. Most slopes are rugged and cut by large canyons and valleys. Sediments are less well known than on the shelves, but are generally 60% mud, 25% sand, 10% rock and 5% ooze and shells. The sediment shear-strength is generally low, 0.1 up to 15 psi (0.7–103.4kPA) with an ocean average of 1 psi (6.9kPA).

In high latitudes where glacial agencies have been at work in the past, significantly large boulder (2–3 m in diameter) and complex steep valleys may have been generated. Even on a global basis, rocks of average size up to 1 m can be expected. Complex and rough terrain will also be present in volcanic regions, where outcropping rocks can cause ridges 2–3 m high. The inclination of the seafloor slope will vary from a minimum of 1° to a maximum of 45° locally.

Seafloor soil characteristics will be vital to any seafloor construction project, especially foundation loading and design. A detailed site survey involving slope measurement, shear strength tests, plate bearing strength and an undersea geological survey of the area will be necessary. Rock debris may be of significant size to require some preliminary clearance of the area. Rock outcrops and ledges may endanger submersible operations. Fine sediment may reduce visibility if disturbed and soft sediment could trap a submersible on the bottom if it sank-in on landing on the seafloor.

TABLE 2.2 Sea states

Douglas Sea Scale (Sea State)		Beaufort Scale (Winds at Sea)	
Code	Waves (height in ft)	Sea Description	Number	Wind (speed in knots)
0	Calm (0)	Sea like a mirror	0	Calm (<1)
1	Smooth (<1)	Ripples; no foam crests	1	Light air (1–3)
2	Slight (1–3)	Small wavelets	2	Light breeze (4–6)
3	Moderate (3–5)	Large wavelets;scattered whitecaps	3	Gentle breeze (7–10)
4	Rough (5–8)	Small waves; numerous whitecaps	4	Moderate breeze (11–16)
4	Rough (5–8)	Moderate waves; some spray; many whitecaps	5	Fresh breeze (17–21)
4	Rough (5–8)	Large waves; more spray; whitecaps everywhere	6	Strong breeze (22–27)

5	Very rough (8–12)	Sea heaps up; foam begins blowing in streaks	7	Moderate gale (28–33)
5	Very rough (8–12)	Moderately high seas; crests break into spendthrift; foam blows in distinct streaks	8	Fresh gale (34–40)
6	High (12–20)	High waves; sea begins to roll; dense streaks of foam; visibility beginning to reduce	9	Strong gale (41–47)
7	Very high (20–40)	Overhanging crests; foam blown in very dense streaks; visibility reduced	10	Whole gale (48–55)
8	Mountainous (>40)	Exceptionally high waves; sea covered with foam patches; visibility much reduced	11	Storm (56–63)
9	Confused	Air filled with foam; sea completely white with driving spray; visibility greatly reduced.	12–17	Hurricane (64–>118)

2.3.8 Seismic and Volcanic Activity

Underwater seismic and volcanic activity is a possibility, especially around the Pacific coast and along the Island Arc. Mapping of epicentres by computers on a world-wide basis is possible, but extrapolation of this work to undersea activity is difficult because of the lack of undersea seismic data. Theoretical studies that have been undertaken on the dynamic response of underwater structures to vertical and lateral forces of excitation suggest that earthquake load may be a major design load for the submersible docking interface and structure base foundation. The underwater structure may also experience more severe loading from low frequency excitation than a similar land based structure.

Earthquakes and volcanic eruptions on the seabed can also create significant waves, which are of the progressive circular type and travel with high velocity through the deep ocean, only appearing as a destructive bore when they reach the continental shelf. These waves are called tsunamis and occur most often in the Pacific Ocean, because of the high seismic activity in that area.

The design of a structure for such active areas will require major design modifications to account for the additional hazards, so operations in such areas should be avoided if possible.

2.3.9 Waves

Generally sea waves are generated by winds. The characteristics of wind-generated waves are determined by the geographical area of origin, the wind velocity, the fetch over which the wind acts and the time duration of the blow. Local sea conditions are a function of the local climate and the geographical location, while swell is generated remote from where their action is observed. Periods of waves generally vary up to 20-30 sec, while longer waves are generated by storm, tides and earthquakes. The earthquake waves are of little practical importance for surface operations as they normally pass undetected subsea.

Wind generated waves occur as air passes over the surface of the water, because a friction or shear force action causes the surface waters to mount up. The waves move in the direction the wind blows and continue in that direction even if the wind direction changes; they are attenuated little on their passage (Table 2.2).

The highest wave recorded was of 37 m, but even in the stormy North Atlantic, waves rarely reach heights in excess of 15 m. Generally a wave will break if its height exceeds a seventh of its wavelength. A wave essentially consists of water particles in a circular motion with a small progressive motion. As the diameter of this motion reduces with water depth, the subsea action of a wave reduces to virtually nothing at depths approaching half a wavelength (Fig. 2.6).

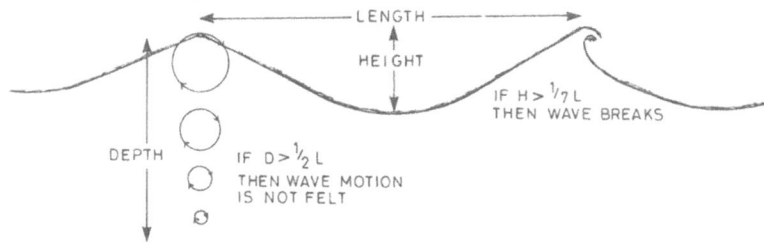

Figure 2.6 Wave movement (Ref. 3)

Waves of sufficient magnitude can be of major concern to surface ocean support tasks. Support vessels while rolling and pitching in seaway, act like pendulums. If the impressed force impacted by the waves becomes synchronous with vessel motion, amplitudes increase significantly. All surface support operations will be complicated by adverse surface conditions and in some cases may need to be terminated. Launch and retrieval of submersibles may be limited to relatively minor sea conditions, and it may be difficult to locate a submersible on the surface because of its low freeboard. Tanker loading systems and floating service platforms will respond dynamically to waves, and in severe conditions, these larger structures may also need to terminate operations. It may be possible to utilize the reduction of effective wave motion at depth to advantage by designing a subsea launch and retrieval system for the support submersible, based on a surface piercing structure.

2.3.10 Corrosion

The corrosion of materials in an underwater environment will not be uniform, so one cannot simply add compensatory material to equipment. Corrosion protection in the totally underwater environment is more easily achieved than at the water line. The type of protection required will be determined by considering factors such as temperature, pH, pressure, fluid flow and abrasion.

In the simplest form, adequate protection may be provided by paint coatings, although sacrificial anodes and impressed current protection systems are in common use today. Sacrificial anode systems, using zinc, offer short term protection before replacement of the anodes. The anodes are normally placed to protect 'what they can see'. The impressed current system is more versatile and flexible, judicious distribution of non-consumable electrodes will allow the protection of complex structures of large surface area.

The requirement to protect a large subsea structure in deep water from corrosion is within the realms of modern technology. The impressed current system would seem more appropriate, because of its flexibility and the limited maintenance requirement.

2.3.11 Fouling

Different combinations of marine organisms are found in the ocean environment depending on depth. Marine life is more prolific at shallower depths and with increased operating depths the fouling potential decreases, generally due to the lack of light from the surface. The critical area for fouling will be mating hatches, as organisms may affect the sealing surfaces.

Small, soft and brittle organisms caught between the sealing surfaces are likely to be crushed and extruded to an extent that they will not affect the seal. Larger more resistant species, crabs, anemones, etc., may affect the seal but these are more likely to be susceptible to antifouling substances that could be released intermittently from reservoirs. Submersibles using manipulators could also be used to clean hatch seals as required.

Underwater structures, due to the discharge of cooling waters or waste material, may generate their own local ecological system around them, encouraging marine growth and the presence of related scavengers. The implications of such an occurrence to operational requirements and the environment should be considered closely before finalization of system design.

2.3.12 Ice

In the higher latitudes, sea ice, glacier ice and river ice can be significant constituents of the ocean environment. In the Northern Hemisphere, glacier ice as icebergs has been known to reach 48°N, and in the Southern Hemisphere, ice from the Atlantic Shelf can survive drifting to 40°S. River ice is only a threat in coastal areas, and sea ice will only form when the total water column is cooled to freezing temperatures. Ice is a great hindrance to surface support operations and undersea operations, as it changes its form and location. The location of an undersea structure in ice-infested waters is another dimension of study involving iceberg dynamics, ice-scour and other local conditions.

2.3.13 Hostile Sea Life

Fundamentally there are two classes of marine life that are dangerous to man, those that bite and those that sting. Sharks are possibly the greatest threat, but only a few species are true killers and all attacks have been in warm waters. Mammals, such as the sea leopard and killer whale, also pose dangers. Fish such as the barracuda, moray eels, and manta rays should be treated with caution. Venomous fish, sea snakes, jelly fish and urchins can inflict painful stings, which in limited cases can be fatal.

Swordfish have been known to attack submersibles and they could cause a catastrophic accident if penetrations or entanglements occurred in sensitive areas. Mooring lines and cables have been known to be bitten and in some cases severed by fish. Large mammals, such as whales, have been known to become entangled on submarine cables and playfully overturn sailing boats!

As man will be enclosed in the structure or protected by the pressure hull of a submersible during transit, there should not be a direct interface with marine life. Design of subsea pipelines, mooring lines and control lines should consider the possibility of large fish and mammal entanglement and the consequences. Lines should also be periodically inspected for fish 'bite'.

2.3.14 Debris on the Ocean Floor

Many forms of dangerous debris litter the seabed, as, from the beginning of civilization, mankind has been using the sea as a dumping ground for his waste material. Explosives, military ordnance, radioactive substances, scrap cables, agricultural poisons and wrecks are a few of the dangers that await the unsuspecting submariner. Explosives are the greatest danger, as mines, shells and torpedoes can be encountered in an unstable state, even outside controlled dumping areas. Unserviceable cables discharged over the side of ships can lie hidden in sediments or suspended from ridges, and pose special dangers to submersible operations.

A detailed site survey of the location and submersible operating area, combined with a clear-up operation of identified dangerous debris, should minimize the dangers from this source.

2.3.15 Light Transmission

As sunlight enters the ocean it is absorbed rapidly by the water. In its natural state sea water is not very transparent and the absorption of the light occurs as a function of colour and depth. Colours at the red end of the spectrum are absorbed first then orange, yellow, green and blue in that order as the depth increases. In the deep oceans blue penetrates furthest.

At depths greater than 100 m there is insufficient light to support the photosynthesis of plant life and for practical purposes light, visible to the human eye, does not penetrate below about 150 m in clear ocean water. Beyond this depth the ocean is a black void where man must bring his own light sources.

Sea water contains organic and mineral particles in suspension, which will cause light to scatter and reduce the ability to distinguish objects underwater. The turbidity, which is high in the photoplankton layer at the surface, is lower in deep water but will still be present due to microscopic sea animals and organic waste falling to the seafloor. The turbidity below the thermocline is very low, but will be increased if bottom sediments are disturbed.

In deep water under favourable conditions the limit on the visual range may be 70 m or more, but is generally 10 m or less, and in coastal high current regions can easily approach zero.

Artificial lighting will be a definite requirement for the depths of operation. Selection of lights with dominant spectrum in the blue-green region, will give maximum transmittance distances. Turbidity should be minimal beyond 1000 m, provided bottom sediments are not disturbed and the underwater structure does not encourage the growth of suspended marine organisms locally. Turbidity will be more restrictive the closer the structure is to the surface. Viewing distances from the structure, with suitable illumination will probably be limited to 70 m at best.

2.3.16 Sound Transmission

The acoustic properties of sea water are most important for underwater operations, as they provide the primary means of communication between subsea systems and surface support facilities. Electromagnetic radiation cannot be used as it is attenuated very rapidly in water especially at higher frequencies.

Sound travels over four times faster in water than air, 1500 m/sec. The speed will increase with increases in temperature, hydrostatic pressure and salinity. Generally sound waves travel at the same speed regardless of the frequency. Higher frequencies provide better resolution, but are attenuated more than lower frequency waves. Sound waves are reflected and refracted and obey the normal laws of physics.

The sea bottom is an ideal reflector, but because the ocean is not a homogeneous medium, numerous boundary layers and density differences will refract the sound waves. Thermoclines are infamous refractors which break down subsea/surface communications or can be used constructively as acoustic communication channels over thousands of miles. Multiple transmission problems are created by bubbles, organic and inorganic matter suspensions, noise-producing animals and deep scattering layers of small vertically migrating animals.

Acoustic communication and location devices will be a crucial element in the operation of the complex. Multiple systems with built-in redundancy will be required for communication links between surface vessels, the underwater structure and submersibles, and for associated location and detection systems. In deep water the thermocline will provide the major barrier to acoustic communication with the surface, contact may be lost with a submersible or structure at depth, or in other cases the sound signal received will give an erroneous reading. Proper design and relatively localized operations should allow a suitable configuration to be implemented to minimize discrepancies and provide a reliable operational system. Standardization of operating frequencies for tasks should be implemented to tie in with international communication standards.

3

OPERATIONAL REQUIREMENTS
AND SYSTEM CONSTRAINTS

3.1 INTRODUCTION

The specification of suitable surface and subsea systems to satisfy
the logistic support requirements of a manned underwater structure in hydrospace, requires initially an analysis of the operational
requirements of such systems and the constraints that may be
imposed. Candidate systems that could provide the required services are then selected, and major limitations of such operating
systems identified. This chapter attempts such a specification of
logistic support before proceeding to a more thorough examina-
tion of critical operational areas.

3.2 OPERATIONAL PARAMETERS

Basic operational parameters for system definition are provided in
Table 3.1 and are determined by expected operational areas,
environmental conditions and location.

Once a site of operation has been chosen, these parameters
could be defined in a more specific manner for the local conditions.

TABLE 3.1   Operational parameters: manned subsea production complex

Parameter	Range
Depth	0–300 m transit only 300–2000 m operating, transit
Ambient pressure	101.4kPA–20,270.7kPa (14.7psi–2940psi)
Water density	1031kg/m^3–1036kg/m^3 (64.37–64.67 lb/cu ft.)
Sea state	6
Significant wave height	6 m
Water currents	surface 0–5 knots bottom 0–2 knots
Visibility	0–70 m
Bottom slope	15° maximum
Distance from land	322 km (200 miles)
Support frequency	day to day

3.3 OPERATIONAL REQUIREMENTS

The schedule of operations that will be required to support a manned underwater structure are listed. Major operational areas are defined, each area and its inter-relation with other system design considerations will be analysed in further detail in later chapters.

1. Flotation of the completed underwater structure and foundation in the building basin and a surface tow to the site location offshore.

2. Controlled deployment of the structure and foundation through the water column to the seabed.
3. Deballasting, levelling and anchoring of the structure or foundation to the seafloor.
4. Transport of personnel and materials from the surface interface or from midwater through the water column to the structure.
5. Location of subsea structure by underwater vehicles and a controlled approach to mating surfaces on the structure.
6. Capability of subsea vehicles to dock, mate and perform pump-down to achieve a one-atmosphere transfer of men and materials between the vehicle and the structure.
7. Incorporation of a suitable ingress/egress system design that allows fast, efficient and safe transfer of men, materials and casualties between the vehicle and the structure.
8. Perform transfer of men and materials from the structure to a subsea vehicle and return to the surface or to a mid-water base.
9. Provide a means of rescuing personnel from the structure by the use of underwater vehicles.
10. Perform safe launch and retrieval of underwater vehicles at the surface or from subsea bases in adverse weather conditions.
11. Payloads delivered to the structure for off-loading should be balanced by the payloads returned to the surface or a mid-water location.
12. A secondary means of rescue from the structure should be provided using free buoyant ascent capsules integrated with the structure construction.
13. All systems should function, safely and efficiently, in the specified environmental conditions.
14. All systems should, if possible, function so as to provide day-to-day support to the underwater structure.

3.3.1 Auxiliary Operational Requirements

These requirements are not specific to general support needs, but are important for the safe and efficient operation and maintenance of the underwater complex.

15. Before emplacement of structure, perform detailed site surveys and clear debris off seabottom location and in the operational area.
16. Continuously monitor environmental parameters, currents, temperature, density, salinity etc.
17. Implement communication systems and command structures between structure, subsea vehicles and surface vessels.
18. Provide subsea location aids and guidance systems.
19. Provide systems to perform external (ambient pressure) tasks, i.e. inspection and maintenance of external areas of the structure, subsea wellheads, control lines and flowlines and anchoring systems.
20. Provide means of cleaning accumulated (debris) sediment from around the structure.

3.4 SYSTEM CONSTRAINTS

The performance of any logistic support system will be influenced and restricted by the imposition of many external factors. These factors may complicate operations, limit the successful attainment of tasks and in some circumstances cause the termination of operations. In support operations requiring a set of systems to work in sequence, the failure of the 'weak link' will cause total system failure. The limiting factors of concern in the logistic support of an underwater structure are now investigated and can be classed into four major categories:

1. Environmental Constraints.
2. Support System Constraints.
3. Production System Constraints.
4. Human Factor Constraints.

3.4.1 Environmental Constraints

In the three-dimensional concept of oceanographic operations, environmental constraints will operate in three areas:

(a) At the air/surface interface.

(b) In the water column from the surface to the seafloor.
(c) At the water/seafloor interface.

(a) *Air/Surface Interface*

Constantly changing weather conditions with the associated change in sea conditions, winds and surface currents will affect the following areas of operation:

(i) Operational planning for float-out from basin, tow-out and emplacement of the structure at location.
(ii) Deployment of the complex through the water column.
(iii) Continuity of service from underwater vehicles transporting personnel and materials to and from the structure.
(iv) Capability to rescue personnel from the structure in an emergency.
(v) Stability of surface support vessels and platforms.
(vi) Tanker loading operations of field products.
(vii) Deployment of surface vessels from home port and aircraft movements.

(b) *Water Column Constraints*

Characteristics of the water column, i.e. density, currents and temperatures will affect subsea operations of equipment and support vehicles.

(i) Water depth determines
(a) Pressure hull design of the structure, submersibles and escape capsules.
(b) Pressure suit design, one-atmosphere diving suits.
(c) Design of encapsulated subsea operational and production equipment.
(d) Operational role of autonomous submarine; can it mate with structure or must it hover at an intermediate depth?
(e) Transit times for ascent/descent phase for the underwater support vehicle.
(f) Location and tracking requirements for the underwater vehicle.

(ii) Environmental characteristics
 (a) Subsea currents will affect underwater vehicle operations, control cables, flowlines, risers and surface mooring systems.
 (b) Variations in density are critical to systems dependent on buoyancy characteristics for operation, i.e. vehicles, structure deployment.
 (c) Turbidity will restrict viewing.
 (d) Hostile agents and debris pose threats.
 (e) Corrosion of the operating systems.

(c) *Water/Seafloor Interface*

 (i) Physical characteristics of the seafloor will determine
  (a) Foundation design for the structure.
  (b) Pipeline and control cable laying techniques.
  (c) Anchoring of surface piercing structures.
 (ii) Topography of seafloor will affect
  (a) Location of underwater structure by underwater vehicles, their direction of approach and safety of operations.
  (b) Maintenance and inspection operations on the outside of the structure and work on associated equipment in the operational area.
  (c) Line of sight of seafloor communication systems.
  (d) The possibility of turbidity currents occurring at location.
  (e) Current distribution by the creation of eddies etc.
 (iii) Debris on the seafloor may affect the safety of the structure and the submersible operations.
 (iv) Local ecology of the structure may encourage marine growth and attendant scavenger sea-life.

3.4.2 Support System Constraints

These constraints are imparted to the overall system by the limitations in the operational capabilities of the sub-systems due to their restricted operating ranges. These systems can be evaluated in three similar groups:

  (a) Surface Support and Transport.
  (b) Subsea Support and Transport.
  (c) On-Bottom Activities.

(a) *Surface Support and Transport*

   (i) Transit times from base to site location.
  (ii) Weather window for launch/retrieval of the underwater vehicles and deployment of the structure.
 (iii) Weather window for tanker loading operations.
  (iv) Limited buffer storage at loading platform for production operations.
   (v) Limited crew accommodation and support services at the surface.
  (iv) Anchoring systems may pose a danger to subsea systems.
 (vii) High cost of surface vessels on stand-by duties.

(b) *Subsea Support and Transport*

   (i) Restricted operating depths of the underwater vehicles.
  (ii) Limited manoeuvrability of the submersibles and bells fitted with thrusters.
 (iii) Restricted payload of the undersea vehicle.
  (iv) Restricted endurance of free-swimming underwater vehicles.
   (v) Restricted dimensions for the carriage of goods and personnel, i.e. hatch openings and limited length allowed.
  (vi) Pressure spheres will not allow maximum space utilization.
 (vii) Buoyancy of the undersea vehicle will require balanced payloads for descent and ascent phases.
(viii) Requirement for rescue capability.
  (ix) Transit time for ascent/descent.
   (x) Communication to surface vessels and structure.

(c) *On-Bottom Activities*

   (i) Limited payload of undersea vehicles for the rescue function.

   (ii) Lack of manoeuvrability for the docking and mating procedure.

  (iii) Ingress/egress, lock-in and lock-out restrictions.

  (iv) Limited maintenance tasks that can be carried out by the underwater vehicles on external equipment.

   (v) Deployment of one-atmosphere diving suits to remote operational areas for maintenance tasks.

  (vi) Refuelling requirement for the subsea power sources.

 (vii) Limited communications due to topography.

3.4.3 Production System Constraints

The use of a manned underwater structure for the production of oil and gas in deep waters, will involve additional constraints to the overall logistic support system design. They will, however, be generally unique to this application of an underwater structure. Some of the factors that should be considered are:

1. Production from wells will be at high pressure and temperature.
2. Presence of a multiplicity of electrical/electrohydraulic control cables lying across the seafloor to satellite wells and manifolds.
3. Presence of flowlines across the seafloor to manifolds and satellite wells.
4. Requirement to maintain control lines and flowlines, wellheads and manifolds.
5. Use of risers to bring products to the surface.
6. Need for the use of a rig for workover operations.
7. Falling debris from the surface operations may create hazards.
8. Avoidance of environmental contamination due to operations.
9. Fire hazards may exist or atmospheres may be contaminated.
10. Continuous production should be maintained.

3.4.4 Human Factor Constraints

1. Personnel must be assured of the safety of the structure, support systems, survival and escape systems.

2. Satisfactory environmental control and comfort in under-water vehicles and escape capsules.
3. Acceptance of the equipment function and design.
4. Suitable duty˙ periods for the personnel within the instal-lation.
5. Skill demands and training requirement.
6. Acceptance of one-atmosphere suits for external seafloor tasks.
7. Surface accommodation.

3.5 CANDIDATE LOGISTIC SUPPORT SYSTEMS

The selection of suitable ocean technology systems to match the logistic support required for the underwater complex, can only be based on presently available, or extrapolated, capabilities of existing ocean systems. A breakdown of such systems available is now considered. These can again be split into the familiar three areas.

3.5.1 Surface Support Vessels

(a) Various ocean-going tugs and barges for towing out the structure and the emplacement phase.
(b) Vessel to provide direct surface support to the structure and launch and recovery of the underwater vehicles, (i.e. Semi-Submersible Service Platform).
(c) Vessel to provide workover operations to the wellhead equipment and launch and recovery of the underwater vehicles (dynamically positioned semi-submersible).
(d) Vessel to act as a storage tanker and perform launch and recovery of the underwater vehicles.
(e) Surface piercing loading tower for export of oil production with a ramp for subsea launch and recovery of the under-water vehicles.
(f) SWATH vessel acting as transit vessel, launch and recovery platform for underwater vehicle operations and a workover platform.

3.5.2 Mid-water Vehicles

(a) Free-swimming submersible of sufficient payload, operating depth, endurance and manoeuvrability for the tasks and with a one-atmosphere transfer capability of men and materials.

(b) Tethered submersible of sufficient payload, operating depth and manoeuvrability for the tasks and with a one-atmosphere transfer capability of men and materials.

(c) Submarine bell of sufficient payload, operating depth and manoeuvrability (assisted by thrusters and a hauldown cable) with a one-atmosphere transfer capability of men and materials.

(d) Autonomous submarine, capable of travelling submerged from base to site location and back to base, with the capabilty to dock to the structure and perform one-atmosphere transfers of men and materials.

(e) Autonomous submarine, capable of travelling from base to site location and back to base with a submersible locked onto the hull. The submersible to have the diving depth capability of the structure, the submarine a mid-water depth capability with hover controls. Submersible must be capable of one-atmosphere transfer of materials and men between the submarine and the structure.

(f) Autonomous submarine with mid-water capability and in the mothership role, capable of stand-by periods on the surface.

(g) Autonomous submarine with a diving depth capability of the structure acting in mothership role, with the capability for stand-by on the seabed.

3.5.3 Sea-Bottom Systems

(a) One-atmosphere buoyant ascent capsules for group escape from the structure, capable of depth of operation of the structure and with suitable surface seakeeping characteristics.

(b) Submersibles as before but acting in the rescue role.

(c) Bell as before but acting in the rescue role.

(d) Submersibles to perform subsea inspection, maintenance and deployment of the seafloor equipment and one-atmosphere suits to remote locations.

(e) Remote controlled vehicle for inspection and minor maintenance.

(f) One-atmosphere diving systems for external maintnance of the structure and more difficult maintenance tasks on remote equipment.

(g) Autonomous submarine of the operational depth of the structure capable of locking-out one-atmosphere diving systems.

3.6 CRITICAL OPERATIONAL AREAS

The previous analysis indicates that several critical areas exist in the provision of logistic support for the structure, these are now identified and will be investigated in detail:

1. Transport of the large structure from the building basin to the site location by towing on the surface, and deployment of the structure through the water column to the seafloor.

2. The dynamic characteristics of the surface support platforms (Semi-Submersibles) and vessels in varying sea conditions will crucially affect the launch and recovery techniques of the underwater vehicles (submersibles and bells), tanker loading operations and other services supplied from the surface. Anchoring systems may also produce hazards to the structure and submersible operations.

3. Docking and mating operations of the submersibles to the structure, reliably and safely is necessary. These operations will be even more critical if an autonomous submarine is used, because of the potential for disaster if a collision should occur.

4. The provision of a suitable ingress/egress system from the submersible/bell to the underwater structure will be necessary to minimize transfer times of men and materials and to retain the integrity of the pressure hull of the structure.

5. Careful attention will need to be paid to the design of the

external equipment so that maintenance requirements match the tool packages of submersibles, remote-controlled vehicles and one-atmosphere diving systems.

6. Topography will affect seafloor communications and mid-water characteristics, submersible to surface and structure to surface communications.

4

TRANSIT AND
EMPLACEMENT
OF STRUCTURES

4.1 INTRODUCTION

It is envisaged that the underwater complex will be constructed by
the combination of five end-capped horizontal cylinders fixed to a
caisson type foundation base. The total system will be assembled
and tested in a dry basin on a convenient shoreline. The complex
will then be secured, the dock flooded and the total system floated
out from the basin and towed on the surface to the prepared site
location offshore. The structure will then be deployed from the
surface, through the water column to the seabed and secured to
the seafloor.

The movement of such a large structure will be restricted by
many factors. Surface conditions will affect tow speeds, loading of
the structure and the operation of surface support vessels during
emplacement. The deployment of the structure through the water
column will require strict control of buoyancy and positioning and
control of displacement caused by subsea currents.

Movements of the structure will only be undertaken when suitable weather conditions appear to dominate, however due to the
relatively long term nature of such operations, adverse conditions
may occur during the towing phase and to a more limited extent
during emplacement. The system design should therefore consider
environmental parameters and hydrodynamic factors that will
require suitable safety design features to be incorporated.

4.2 SITE PREPARATION AND ASSESSMENT

A detailed site survey will be required prior to the station design finalization and emplacement. Environmental conditions at the site and over the transit route should be studied and incorporated into the design. Distances from land based support areas should be evaluated for logistic support requirements. Available surface support vessels, i.e. ocean tugs, barges and submersible support vessels for the task requirements, should be assessed.

Subsea survey of the site location could be carried out by a special work submersible equipped with a tool package required to measure the site parameters. Environmental data, i.e. temperature, pH, salinity and currents, need to be collected. The seafloor will have to be surveyed for potential dangers to the structure, such as underwater debris, and the topography mapped. Seafloor soil characteristics, sediment cover, plate bearing strengths and cores will be required before the foundation design is finalized. The slope of the seafloor will also be critical to the foundation and emplacement requirements. Geological features of the operational area will need to be studied and the seismicity and vulcanology of the area evaluated.

The bearing strength of seafloor materials will determine the maximum weight of the structure that is acceptable, and the most suitable foundation design. Fine grain sediment in the area may greatly hamper visibility and seismic considerations may determine limiting design criteria in some sub-systems.

Generally, the instrumentation exists to measure most of the parameters *in situ*, although a suitable set of submersible tools may need to be designed to perform the measurements. Exploration drilling operations will be undertaken in the area prior to development of the field, and a considerable amount of relevant data could be collected during this period for the initial design studies of the structure. Clearance of debris and hazards identified at the site could proceed after well completion, while a semi-submersible drilling rig is available for lifting operations.

4.3 TOWING

The collection of cylindrical modules forming the seabed facility will be located on and fixed to a caisson, which will form the foundation on the seafloor at the site location. The structure is near neutrally buoyant for the emplacement phase and will normally float with 75–100% of the structure submerged. The caisson will provide sufficient buoyancy and freeboard to give the necessary hydrodynamic characteristics for a safe and efficient tow of the structure on the surface.

There will be several limitations to towing operations:

1. Minimum draught.
2. Width.
3. Hydrodynamic stability of surface movement and drag effects.
4. Induced loads on the structure due to dynamic wave loading.
5. Limited towing speeds.

4.3.1 Draught and Width

The draught of the structure will impose limitations on the worldwide deployment of the system from a major building base. The incorporation of suitable buoyancy in the foundation caisson can reduce the draught, however it is obvious that the draught requirements envisaged will be limiting.

TABLE 4.3 D.S.P.S. Consortium five module complex (module dimensions 12 m (I.D.) × 68 m long)

Depth (m)	Construction	Foundation	Base Dimensions (m)	Draught (m)
200	Reinforced concrete	Gravity caisson	66 × 87	18.6
500	Reinforced concrete	Gravity caisson	112 × 82	22.5
1000	Steel/ concrete composite	Steel template piled	66 × 87	14.1

Port facilities generally have a draught of the order of 10-12m, although since the increase in the use of super tankers, deeper facilities are now more generally available, i.e. Europort, Rotterdam at 22 m. A safety factor of 2-3 m should be allowed below the summer salt water draught of large vessels to account for density changes, squat, pitching and rolling.

The draught restriction will limit the available shore bases suitable as an assembly basin and will determine the towing distances necessary to deploy the structure to the site location. Worldwide transit by the normal canal routes of the Panama and Suez will not be possible, as their limiting depths are 12 and 13 m, respectively. A five module complex will also exhibit a width of 80 m, and would not allow passage: the Panama Canal maximum width being 33.5 m. The towing of such large and expensive structures between oceans by routes such as Cape Horn, is not advisable because of the frequent occurrence of severe gales. Oceanic towing distances should also be minimized where possible to reduce the risk to the structure from adverse weather conditions.

These major restrictions due to large dimensions and hydrodynamic considerations will minimize available construction sites and deployment possibilities. Construction will probably be undertaken in the ocean of operations, and limited port facilities will probably involve long ocean tows to the site location.

4.3.2 Hydrodynamic Considerations

The structure, when on tow on the surface of the ocean, should exhibit stability. To achieve this, suitable buoyancy adjustments must be possible and an effective towing arrangement must be devised. It may be necessary to incorporate a fairing structure to minimize the effects of hydrodynamic drag on the structure. Tow tank tests of hydrodynamic models will be required to determine the towing characteristics of the structure and to allow the design of a suitable configuration.

The complex and foundation during tow will be exposed to stresses imposed by dynamic wave loading that it will not normally experience on station. Temporary or permanent reinforcement of the structure during tow may be required to counteract such forces and prevent damage to the structure in bad weather conditions.

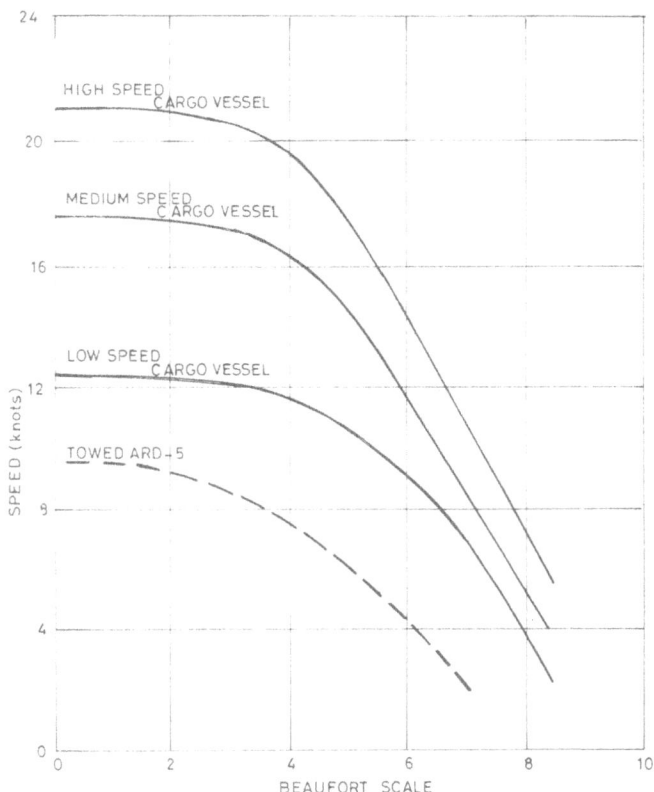

Figure 4.1 Average transport speed loss in head seas

Submarine towing of such a large structure is not practical, because of difficulties of buoyancy control.

4.3.3 Towing Speed

The speed at which the structure can be towed will be determined by several factors:

1. Hydrodynamic characteristics of the structure on the surface.
2. Prevailing sea-state and wind conditions.
3. Capability of the towing vessels.

Generally, rough weather will reduce towing speeds due to

voluntary and forced reductions in speeds. Forced reductions are related to an increase in resistance due to rough weather and reduced propellor efficiency of towing vessels from loading, air draining or reduced hull efficiency. Voluntary reductions are made to prevent damage to the ship, loss of tow or danger to the crew.

Experience of ocean tows indicates that below Beaufort wind values of 4, the average decrease in towing speed per unit increase on Beaufort scales is about ½ knot. Above Beaufort 4, towing speed reduction is normally about 2 knots for each unit increase (Fig. 4.1). Average towing speed, based on Beaufort 4 is about 6 knots, although current vectors will affect average speeds.

Towing of large offshore production platforms occurs at 2–3 knots. For this structure, with a suitable fairing to reduce hydro-dynamic drag, we could expect towing speeds between 2 and 5 knots.

4.4 EMPLACEMENT AT SITE LOCATION

The deployment of the structure from the surface at the site location through the water column to the seabed will be the most critical phase of operations. Surface support operations will need to be undertaken in calm weather conditions if the structure is suspended from cables to surface vessels, as pendulum effects will adversely affect operations.

For such large structures the use of controlled buoyancy is likely to be the major method of deployment of the structure subsea. Ideally, such a large structure should gain buoyancy as it descends, requiring less operational control. Depth (pressure), the bulk modules of the hull, seawater and temperature changes are the most significant factors in determining buoyancy changes, and will need to be measured at the site prior to deployment. A detailed consideration of how the weight to buoyancy relationship changes with depth is required, as well as what descent velocities can be achieved safely to avoid excessive drift and descent times, and minimize impact with the seabed. The dynamic stability of the structure while travelling through the water column, may also be of concern (rotation and oscillation) if descent speeds are sufficiently high.

For a structure that gains buoyancy with depth to be neutrally buoyant at depth, it will need to be heavy at the surface by an amount equal to the buoyancy gain at depth. The excess weight at the surface will provide the downward velocity, increases in speed can be achieved by additional weight. As the station goes deeper, speed will reduce with gain in buoyancy and, if the additional weight touches bottom before the structure, this will assist the deceleration phase and leave the structure neutrally buoyant just above the seabed. Final ballasting can be used to lower the structure to the seabed.

Deployment of the structure from the surface to the seabed should be achieved as quickly as possible because the descent is the most critical phase. However slow deployment speeds will only be possible with such a large structure. A structure being propelled by buoyancy forces through the water column will fundamentally be subjected to inertia, gravity and fluid forces. If the negative buoyancy is small, vertical deployment speeds will be small and hydrodynamic forces will be negligible compared to gravity and buoyancy forces. The body will be stable if the centre of gravity is below the centre of buoyancy. Current drag forces may, however, cause rotation of the structure during descent. This could be controlled by structure propulsion thrusters or, with more difficulty, by guide cables.

Incorrect assumption of seawater changes with depth and the appearance of strong currents will produce dangerous situations. Excessive descent velocities may cause instabilities and oscillations and heavy impact on the seafloor. Lack of information regarding the seafloor characteristics may cause encounters with plateaux and ridges or soils of insufficient bearing strength. All factors could cause a catastrophic failure of the system.

Auxiliary systems will have to be incorporated in the complex to provide main ballast and sufficient freeboard for surface operations, with diving and velocity control weight and other payloads attached. Flooding the main ballast should provide negative ballast for the dive. A trim ballast system will be required to adjust the trim on the surface and subsea. Temporary strengthening provided for the structure during tow, will need to be removed before the dive. A submersible should be available for monitoring the subsea installation phase.

As the structure is essentially passive during emplacement, i.e. lacks any major manoeuvring capability, the deployment to a correct location on the seafloor will be determined by the accurate positioning of the surface support vessels guiding the structure. Communications and detailed areas of responsibility will be key considerations in the safe and effective emplacement of the structure. Once on the seafloor the gravity foundation will need to be levelled or the template piled to the seafloor. Ancillary systems connections will then be made and inspections and tests carried out. The crew will then be transferred from the surface to the structure in a submersible.

An initial thorough inspection should be done after thirty days to assess any bad effects from corrosion or stress, with accumulated experience allowing extension to one or two years between major inspections.

4.4.1 Dynamic Response of Emplaced Structure to Currents and Seismic Forces

Surface currents will affect the structure on the surface and produce sheer and rotational forces during the deployment phase. However, due to the large structural weight and negatively ballasted condition of the structure on the seafloor, subsea currents should not affect the stability of the structure. They may, however, cause sediment to build up around the structure and this may need to be cleared periodically.

The possibility of seismic disturbances occurring at the site location will have a direct effect on structure design. Ground motion during earthquakes has a vertical and horizontal component, the vertical component normally being ignored in land studies. Previous studies of the effect of seismic disturbances on underwater structures, indicate that the vertical component may present a more serious problem than the horizontal component. If underwater earthquakes approach the intensity and duration reported on land (highest − El Centro, California, May 1940, acceleration 0.33 g, velocity 13.7 in/sec (0.366 m/sec) displacement 8.3 in (210.8 mm)) a brief over-pressure, higher than the total hydrostatic pressure, might occur that cannot be ignored for its effect on the underwater structure.

The studies also indicated that underwater structures would experience more severe loading from low frequency excitation than land structures and less from high frequencies.

To provide an overall dynamic loading evaluation, seismic loading should be considered in association with hydrodynamic loading data from currents etc.

LOGISTIC
SUPPORT LOADS

5.1 INTRODUCTION

The logistic support of a manned underwater structure through the water column will be a relatively expensive operation in relation to the life time cost of the structure itself and the use of alternative systems such as umbilicals. The aim of support system design, should therefore be to attempt to minimize the cost of these operations while retaining an acceptable level of support in an efficient and safe manner.

Factors affecting the cost of support operations should be assessed. The design of the underwater structure will directly affect the logistic payload requirements, and determine the frequency of support operations. The logistic loads that have to be transported to and from the complex need to be evaluated, and the support vehicle and surface vessel's requirements identified.

Once these parameters have been considered, an operating philosophy for support operations should be investigated that should maximize the utilization of the support systems and provide an acceptable level of service to the underwater structure.

5.2 COSTS OF SUPPORT OPERATIONS

The operational costs of an underwater production complex are likely to be heavily influenced by personnel related costs. A large fraction of these personnel related costs will be in elevator charges for the transfer of men and materials through the water column. The costs will include charges for the movement of the personnel themselves and their food, waste and life support needs. The use of underwater vehicles, with their attendant support vessels for these purposes, will be a relatively expensive procedure without much leeway for cost relief.

Fundamentally, underwater vehicles are expensive transport systems. Charges for the use of such vehicles will primarily be determined by the depth of operation of the vehicle, with a fixed charge generally allotted on a per seat basis. Surface support vessels required for the launch/retrieval of the vehicle and as a standby vessel for the otherwise unattended structure will generate costs that will be primarily size dependent. In essence a high cost-per-volume vehicle, with its associated support costs, will be used to support a relatively low cost-per-unit volume underwater structure. These cost differences will be accentuated with the use of larger submersibles and larger support vessels and with the possible requirement for a second submersible on standby duties for emergency tasks.

Vehicle costs are essentially determined by operating depth and the size of support vessel required to support its operation. Operating depth will be a mandatory requirement, so cost relief can really only be obtained by reducing vehicle size and thus the support vessel size, and maximizing utilization of the support system. Vehicle size will be determined by payload requirements and the number of delivery cycles (dives) acceptable for support operations. Payload will vary with manning level, the inherent design of the complex and the possible use of alternative re-supply methods for materials. Delivery cycles will be a function of ingress/egress system design and safety requirements.

5.3 LIBERAL DESIGN OF THE COMPLEX

The design of the associated operating systems of the underwater complex will directly affect the logistic payload requirement to support the structure and the frequency of supply operations. Previous studies of manned underwater structures have shown that any incremental costs involved in the provision of a higher level of sophistication in sub-system design, is unlikely to affect the total system costs significantly, and that savings in the cost of support operations are likely to compensate more than adequately for these incremental costs.[7]

To illustrate this philosophy, if we consider a basic design of a manned underwater structure that relies on the provision of carbon dioxide absorbent, oxygen candles and water for life support, these items represent a substantial logistic load for supply and removal. A regenerative life support system providing regenerative absorption, oxygen generation and desalination of water, would restrict re-supply requirements to such minor items as filters etc. The initial investment in the life support system may be larger; however, in terms of the life time of the complex, savings in logistic support costs will more than compensate for this additional cost. Also, the frequency of support operations will not be so critical, as the complex design will be inherently safer.

The allotment of a minimum volume for the complex design is also not recommended. The incorporation of a little extra volume in the overall design would not involve any total system cost barriers as the investment involved would again be relatively minor. The additional volume available in the complex could be used advantageously to alleviate support requirements. Food supplies and other items could be stored to support operations in excess of the normal duty cycle period adding more flexibility to supply operations. Intensive support operations could be used to stock the complex at convenient operational times, i.e. in good weather and during maintenance periods. The additional volume would also be useful for the processing and storage of waste prior to removal operations.

Liberal design can also be considered in the provision of power to the complex. At the power levels envisaged, it soon becomes clear that, if an umbilical power supply is unavailable, nuclear power systems are the requirement. The specific costs at high power levels are distributed over a long enough supply life time to bring the costs per kWh down to equivalent mobile units on land, and even at lower power levels costs are still low enough to be tolerated. If considerable power is required for desalination, life support systems and overboard pumping of wastes to reduce logistic loads, the case is strengthened. The great advantage of the use of nuclear power systems is that there is an absence of the continuous demand for the logistic support required by other systems, i.e. batteries, fuel cells etc.

The incorporation of a docking lock into the complex, for the acceptance of naturally buoyant material capsules transported by remote controlled vehicles or submarines, may also be justified if vehicle logistic loads can be reduced to personnel transport only, providing this is not compromised by under-utilization of the submersible.

The incorporation of liberal design features in the complex will have a significant effect on the requirements for, and the cost of, logistic support operations. The incremental costs for such liberal design features will not be excessive in terms of total system cost over the life time of the structure and should be implemented to reduce logistic support demands.

5.4 LOGISTIC PAYLOADS

On the assumption that the underwater structure will incorporate regenerative life support systems, desalination of seawater, a nuclear power system and perform overboard pumping of sanitary wastes, logistic support payloads should be restricted to the movement of personnel, supply of food and other consumable materials and the removal of waste materials.

5.4.1 Personnel

The underwater production complex will have a crew of thirty personnel on duty for a two-week period. To provide operating continuity it is probable that half the crew compliment will be replaced each week, after initialization of the system. The logistic payload for personnel movement will therefore be 3324 kg (110.8 kg × 30) every two week period. The allotment of 110.8 kg for manning, also includes an allowance for baggage and recreational materials as well as the person's weight. The weekly payload if half the crew are transferred will be 1662 kg.

5.4.2 Food

The nutritional requirements for men living and working in the manned underwater structure will be essentially the same as those performing similar tasks on the surface. Dietary requirements will be approximately 2800–3000 cals per day and the food available for consumption should be equivalent to a normal diet. A number of basic food types can be considered for transport which will provide a suitable variety for a palatable diet, e.g. dehydrated, freeze-dried, frozen and conventionally packaged food.

Food management will be critical to logistic support operations as the food requirements on a man-day basis are essentially fixed and accumulative weight and storage requirements cannot be altered significantly. Waste material to be removed from the structure will be a function of the type of food and its packaging requirements, however compaction of wet and dry garbage may be possible to reduce volume requirements. The weight and volume of foodstuffs and storage equipment for thirty men for a fourteen day duty cycle are shown in Table 5.1.

TABLE 5.1   Weight and volume requirements for foodstuff and storage: 30 men, 14 day duty cycle.[7]

Operational Period		Weight (kg)		Volume (m³)	
		per man-day	Total	per man-day	Total
420 man-days	Food	1.30	546	0.0059	2.48
	Storage	1.70	714	0.0065	2.73
	Total	3.00	1260	0.0124	5.21

Total food payload requirement is therefore 1260 kg (2768 lb) with a volume requirement of 5.21 m³ (184.8 ft³), which is equivalent to 90.5 kg (200 lb)/man/month requirement including wrapping and insulation for refrigerated materials.

The logistic support loads will obviously not be restricted to the provision of personnel and their associated food requirements. Other equipment and materials required for the operation of the subsystems within the complex will need to be transported to the structure, i.e. spare parts, filters, lubricating oils, chemicals. A detailed assessment of this requirement can only be undertaken once a system design has been finalized. To account for the use of such consumables, a factor of 1.5 will be allowed against the food load. A total logistic load per man-month will therefore become 135.8 kg (300 lb).

Waste materials to be removed from the structure are more difficult to estimate. If we assume that the complex design will allow the pumping of sanitary wastes overboard, this will reduce removal loads significantly. Wet and dry garbage from food consumption, after compaction, will require transport along with bilge oil and other consumables used within the complex. An attempt is made in Table 5.2 to estimate the likely re-supply and removal loads that can be expected in our concept of an underwater production complex.

The difference in payload for supply (5205 kg) and removal (4680 kg) is due to the fact that we have assumed that sanitary wastes will be disposed of to sea. This effectively extracts the water content of the food and human consumption waste from the removal load. The load difference can be used constructively for removal operations, or easily compensated for by taking on ballast to retain neutrally buoyant conditions in the underwater vehicle.

Table 5.3 identifies the major characteristics of present submersible designs. Payload ranges are shown because the dry payload capacity of the vehicle will vary as a function of the operational equipment fitted to perform the task requirements. The fitting of the transfer skirt and associated equipment to perform dry transfer operations will itself significantly affect the dry payload capacity.

It is clear that even the larger submersibles presently available could not carry the logistic loads envisaged in one dive to the complex, ignoring the complications and difficulties experienced in the launching and recovery of such large vehicles. A multiple-dive supply cycle will obviously be required.

The configuration of the available volume within the submersible hull should also be considered as it is unlikely that full utilization of the space will be possible, and a correction for the packing density of materials and personnel should be allowed

TABLE 5.2 Resupply/removal Loads: 30 Men for 14 Days

Resupply	Weight (kg)	Volume (m³)	Remarks
Manning and Effects	3324	15.56	30 × 110.8 kg, includes allowance for baggage, recreational material
Food and containers	1254	5.23	Assumes 3 kg/man-day at 0.012 m³ man-day
Miscellaneous: i.e.	627	0.87	Assumes 1.8 kg/man-day
(Lubricating Oil	452	0.45	Assumes 1000 kg/m³ density
Filters, Spares)	174	0.42	Cartridges, packages etc.
TOTAL	5205	21.66	

Removal	Weight (kg)	Volume (m³)	Remarks
Manning and Effects	3324	15.56	Off-going crew and baggage
Bilge Oil etc.	678	0.68	Allowance for system leakage, clean-up
Wet/Dry Garbage	678	1.73	Assume density of 392 kg/m³
TOTAL	4680	17.97	

TABLE 5.3 Payload characteristics of dry transfer submersibles

Name	Builder	Operator	Depth (m)	Weight (kg)	Payload (kg)
Mermaid IV	Bruker Meerestechnik Germany	A Foreign Navy	600	17 000	1000
L5	Vickers Slingsby	British Oceanics	475	20 321	180–500
Taurus	International Hydrodynamics Canada	British Oceanics	610	24 000	800–1800
PC 1800	Perry Oceanics USA	British Oceanics	300	11 000	200
PC 16	Perry Oceanics USA	British Oceanics	915	15 000	270
Deep Sea Rescue Vehicle D.S.R.V.	Lockheed Missile & Space	US Navy	1524	34 000	1950
U.R.F.	Kockums Sweden	Swedish Navy	460	49 000	3000

TABLE 5.4 Frequency of support operations and logistic loads

Vehicle Dive Cycle	Payload (kg)		Volume (m³)		Total Payload (kg)	Total Volume (m³)	Remarks
	Personnel	Materials	Personnel	Materials			
One dive every 2 weeks	3324	1881	15.56	6.1	5205	21.66	Autonomous submarine only.
One dive once a week	1662	940	7.78	3.05	2602	10.3	Possible, not recommended.
Two dives a week	831	470	3.89	1.52	1301	5.41	Possible, operational difficulties large submersibles
Three dives a week	554	313	2.59	1.02	876	3.61	Five personnel and materials. Recommended. *1
Four dives a week	415	235	1.94	0.76	650	2.71	Three to four personnel and materials

Personnel and Materials

Personnel Only					
One dive every 2 weeks	3324	—	15.56	3324	15.56	Thirty personnel. Not recommended.
One dive once a week	1662	—	7.78	1667	7.78	Fifteen personnel. Not recommended
Two dives a week	831	—	3.89	833	3.89	7–8 personnel. Recommended *2
Three dives a week	554	—	2.59	556	2.59	5 personnel, too small

when assessing a vehicle for use. A considerable portion of the logistic load is made up of food, materials and waste and alternative methods of transport could be considered for this element, possibly releasing submersibles for personnel transport only.

A breakdown of re-supply logistic loads as a function of multiple-dive cycles and personnel transport only is presented for consideration in Table 5.4.

The various operational cycles shown in the analysis would appear to indicate that a vehicle operating on the basis of three dives a week[1] for personnel and materials transport may be preferred. The payload requirements of 875 kg and volume availability of 3.61 m^3 would not involve a particularly large submersible for the requirement. The numbers of personnel at risk during transfer operations would be acceptable, and personnel exchange for the materials element of the payload (3 men) would provide sufficient capacity for rescue requirements, i.e. 8 men. If the materials payload is delivered by an alternative concept, the vehicle could perform the personnel transfer in two dives per week.[2] The requirement to dive three times per week should allow enough operational leeway to account for restrictions on vehicle deployment due to adverse weather or other factors.

5.5 LOGISTIC SUPPORT OPERATING PHILOSOPHY

The transport of logistic loads from a surface or subsea base to a manned underwater complex will be an expensive operation due to the high cost investment in underwater vehicles and the systems to support their operation. The only way of alleviating these costs is by minimizing support system requirements, the use of alternative lower cost support systems or maximizing utilization.

In our concept of a permanently emplaced underwater production complex, logistic support loads can be significantly reduced by adopting liberal design features in the area of life support, power generation, volume allocation and waste treatment. The additional investment would not be excessive in terms of the lifetime cost of the structure and the support cost savings would more than compensate for the initial investment.

The remaining loads, after the implementation of liberal design, would be restricted to the supply of personnel, food and materials to the complex. The only relief in this area could be obtained by an alternative re-supply method using a lower cost system. One approach may be to consider personnel transport and food and materials as logistically separate items. The latter, without the inherent need for safety considerations, could perhaps be transported in one-atmosphere capsules by lower cost remote controlled vehicles, or by a buoyant tethered capsule arrangement, to a docking lock in the complex.

The operational dive frequency of the underwater vehicle will be determined by submersible payload, support vessel operational limitations and the ingress/egress system design. In general, to minimize the cost of the vehicle and the support vessel and maximize utilization, a small submersible should be used more often. This however, needs to be balanced against the requirements for a payload that gives an acceptable operational sequence and matches the need to carry sufficient personnel during rescue operations from the complex.

It is considered that an underwater vehicle with a dry payload capacity between 750 and 1000 kg and with a transfer hull volume of about 4 m^3 is required. The vehicle will supply the logistic loads to the structure on a multiple-dive cycle of three dives per week. The vehicle and its support systems should not exhibit extortionate costs. The utilization of such a support system should be frequent enough to justify its permanent availability, while the day to day continuity of operations should not be critical to the support of the structure. The operational flexibility of such a concept will probably not require the consideration of alternative re-supply methods for material transport. Ingress and egress system design should not require excessive buffer storage to receive the payload.

6

SUPPORT VEHICLE

6.1 INTRODUCTION

A station support vehicle will be required to transport personnel and materials from a surface support vessel or subsea launching base through the water column to the underwater structure. The vehicle will also be required to provide a rescue capability for the evacuation of personnel from the complex under emergency conditions and to perform as a work vehicle for external maintenance tasks around the structure and within the oilfield area on the seabed.

The operational requirements of the support vehicle in the specified roles need to be defined and an assessment of vehicle characteristics performed. System constraints that will restrict the successful attainment of goals will need to be considered and the total information base correlated to conceive a design of vehicle to meet the task requirements.

Logistic support loads to be carried in transit, the ingress/egress system for transferring loads from the vehicle to the complex and surface and subsea support systems for the vehicle, are considered separately. This section only considers the selection of a suitable subsea support vehicle and associated sub-systems for its effective operation.

6.2 SUBMERSIBLES: GENERAL

Submersibles as a rule are generally designed for minimum weight and maximum performance; features which demand the economic use of the available space on the vehicle. Most vehicles have little volumetric capacity for payload. Gross payload capacity will also have little relation to the dry payload capacity of the vehicle, as much of the payload may be taken up by the operating systems necessary for the efficient performance of tasks by the vehicle.

Many submersibles, especially those capable of deep operation, have a single pressure hull that is generally spherical in shape and with no facility for dry transfer operations. Ideally, a submersible vehicle that is to be used for dry transfer operations, should have two pressure hulls; one for control systems and pilots and the other for transfer operations. The two hulls should be separated by a lock arrangement to avoid loss of the vehicle in the event of a mating seal failure. The transfer hull should also be cylindrical to maximize the available volume for personnel and materials carriage. The incorporation of dry transfer arrangements to existing vehicles requires considerable structural redesign and reworking.

Recent trends, however, in submersible design show an increasing capability for diver lock-out and dry transfer operations, but with an increase in the dry mass of the vehicles. Increases in weight are also a reflection of deeper operating depths, requiring stronger materials and higher battery capacities, and a wider range of instrumentation.

The use of such large vehicles imposes restrictions on the use of launch and retrieval systems and operational capabilities. Surface support operations require large vessels to handle the vehicles of such size and weight. Larger vehicles are also more difficult to control subsea and there is a danger of impacting structures (while undertaking work tasks), especially under fluctuating current conditions.

The major cost of submersible operations is normally determined by the operational costs of the support vessels required for its use. Large submersibles, requiring large support vessels, involve high operating costs, while the use of a small submersible making more dives from a less expensive support vessel, may be more economic. The choice of submersible size will, however, also be a function of the number of transfer cycles that are acceptable for moving the logistic load to the complex.

The dedication of the submersible to the support of a manned underwater structure in the multi-role function as described, will probably require a purpose designed vehicle to match the task requirement. The submersible technology exists today for such a design. The major effort required is the correlation of suitable sub-systems into the design to maximize the efficiency and safety of operations by the submersible in these roles.

6.2.1 Submersibles: General Characteristics

A submersible is primarily a submerged craft and secondarily a surface craft. It should, however, be stable in both modes. A submersible consists of seven major systems:

1. Vehicle Structure.
2. Pressure Hull(s).
3. Ballast System.
4. Propulsion System.
5. Manoeuvring System.
6. Life Support System.
7. Navigation System.

These major systems are basic requirements of the vehicle. The combination of these systems and their capabilities will be determined by the operational requirements. Other systems that are incorporated in the vehicle either augment the vehicle's capabilities or are sub-systems or back-up systems to the major systems.

The operational requirements of the vehicle establish the operational limits of the vehicle and consequently the depth, speed, manoeuvrability, endurance, stability, capacity and navigational requirements for the task performance. The operational role of the submersible has, therefore, to be defined before suitable submersible parameters can be considered.

6.3 OPERATIONAL ROLE OF THE SUBMERSIBLE

6.3.1 Task Requirements (Fig. 6.1)

1. Transport personnel and materials in a one-atmosphere environment, from a surface or sub-sea base, through the water column to the complex and return personnel and waste materials from the complex. Provide sufficient storage for these purposes and achieve reasonable ascent/descent transit times.
2. Locate the underwater complex and make a controlled approach to the docking platform on the complex, avoiding obstacles that may be present in its path.
3. Manoeuvre over the docking plate, land and mate to the complex with minimum impact loading. Perform dewatering of the mating skirt and provide one-atmosphere transfer of personnel and materials to the complex. Undock and return to the surface.
4. Perform the rescue of personnel from the safe haven area in the complex if an emergency situation calls for evacuation.
5. Perform external inspection and maintenance tasks on the complex structure and field operations equipment that cannot be performed by remote controlled vehicles and one-atmosphere diving suits deployed from the complex.
6. Be capable of deploying and supporting the one-atmosphere diving suits and remote controlled vehicles at remote field locations for specialized work tasks that cannot be undertaken by the vehicle's work tools.

6.3.2 System Constraints to Submersible Operations

Chapter 2 considered the constraints imposed on support operations for the underwater structure. These are now *correlated in terms of the specific operational task requirements* and form the basis for vehicle selection parameters.

1. Weather window for the launch and recovery of the submersible from the surface support vessels.
   Limitations on the use of subsea launching and docking systems.
   Restricted operating depths of the vehicle.
   Restricted payload of the vehicle.
   Restricted endurance of the free swimming vehicles.

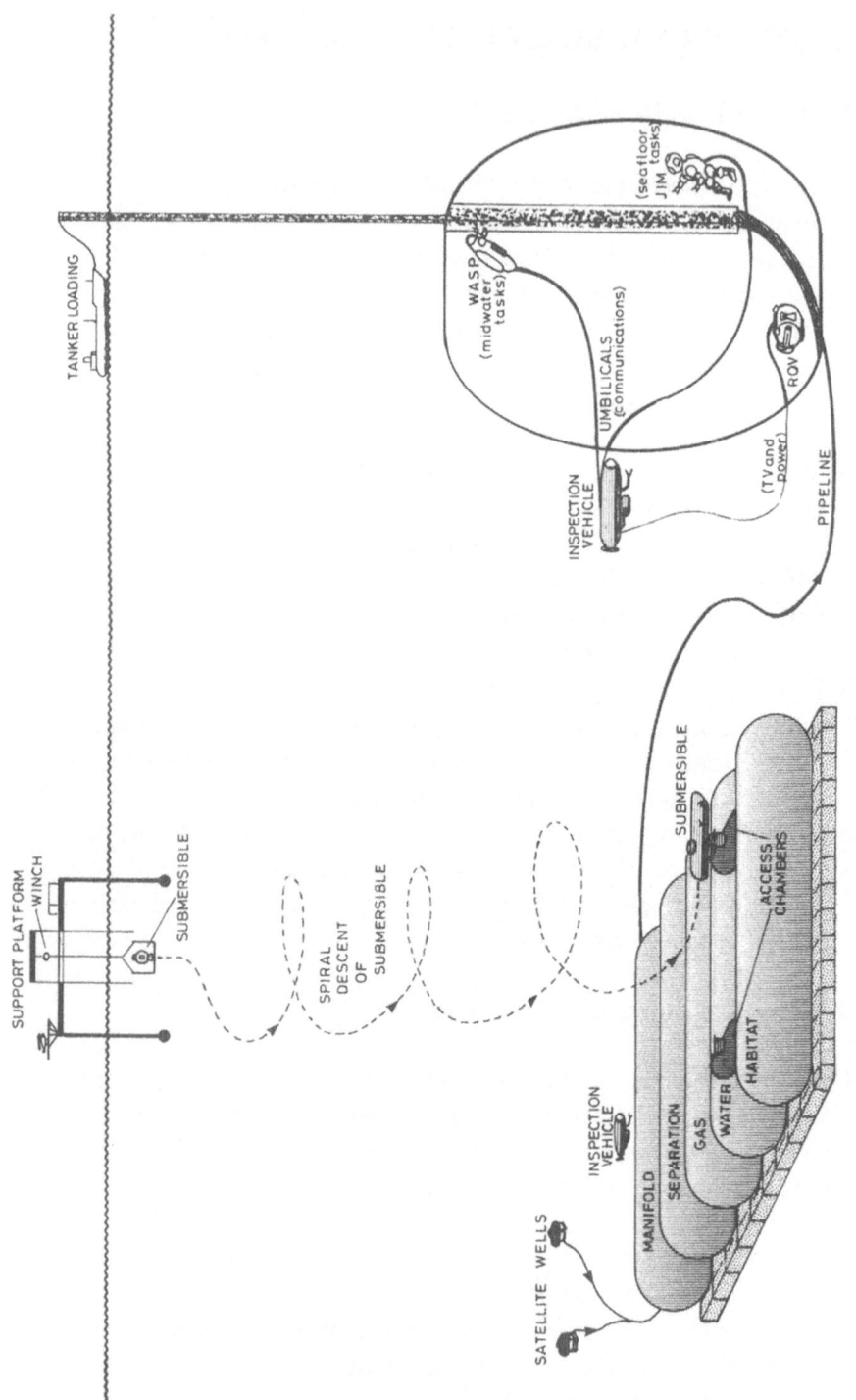

Figure 6.1 Vehicle task requirements

Restricted dimensions for the carriage of personnel and materials.
Requirement for balanced payloads for the ascent and descent phases.
Transit time for ascent and descent.
Communications to the surface vessels and the complex.
Utilization of the payload space in the vehicle.
Presence of control cables and flowlines on the seafloor.
Falling debris from the surface operations.
Life support and comfort of personnel during transit.
Ingress/egress transfer restrictions.
Environmental factors: currents, pressure, density, debris.
Skill demands and training requirements for the vehicle crew.

2. Limited manoeuvrability of the vehicle.
Restricted endurance of the free-swimming vehicle.
Transit time for the ascent and descent phases.
Communications to the surface vessels and the complex.
Environmental factors: topography, currents, turbidity etc.

3. Fitting of the mating skirt will affect the hydrodynamic performance of the vehicle.
Shock mitigation systems or skids may be required.
Power requirement for the dewatering procedure.
Dangers due to impact loading on the complex and the submersible.
Dangers of seal failure when locked onto the structure.
Ingress/egress restrictions.
Buoyancy control during transfer operations.
Limited manoeuvrability of the vehicle.
Hatch mating compatability.
Environmental factors: currents, turbidity, marine fouling, debris, corrosion.
Contaminated environments in the complex.
Skill and training requirement of the vehicle crew.

4. Weather window for the launch and retrieval of the submersible at the surface.
Limitations on the use of subsea launching and docking systems.
Payload, ascent/descent times and communications are more

critical than with normal support operations.

May need to deal with hyperbaric situations in the complex.

Requirement to carry casualties, who may be immobile.

Fire and contamination hazards may be present.

Buoyancy control by balancing load more difficult.

5. Design of subsea operational and production equipment.
   Limitations of viewing capabilities and manipulator performance.
   Debris and operational equipment on the seafloor.
   Lack of manoeuvrability, difficulty of control of the large vehicles.
   Line of sight communications on the seafloor.
   Environmental factors: currents, turbidity, topography etc.
   Skill and training requirements of the vehicle crew.

6. Special attachments for the carriage of one-atmosphere diving suits and remote controlled vehicles.
   Hover requirements.
   Special buoyancy adjustment arrangements for these tasks.
   Umbilical connections to support these systems.
   Increased power requirements for the support systems.
   Endurance, more critical for the support role.
   Environmental factors: currents, turbidity, topography etc.
   Skill and training requirements of the crew.

The system constraints will obviously form the base for detailed submersible design criteria and cannot be evaluated fully in a study of this nature, however, major system elements are considered.

6.4 SUBMERSIBLE FEATURES

The submersible vehicle will be plying from a surface or subsea base to a known location on the seabed. Search operations, which are heavy on endurance, will be of relatively minor duration as the complex will be in a fixed location and various navigation aids will be available. The level of sophistication required for this task will be minimal; however, design complexities will be involved in

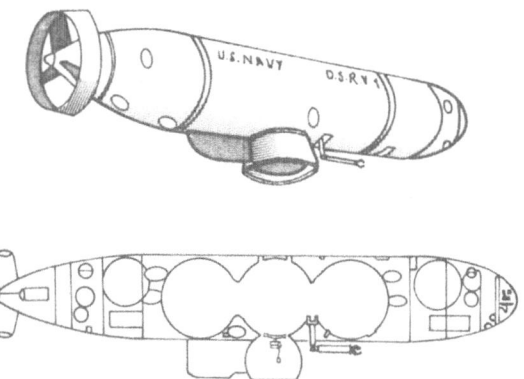

Figure 6.2 D.S.R.V.

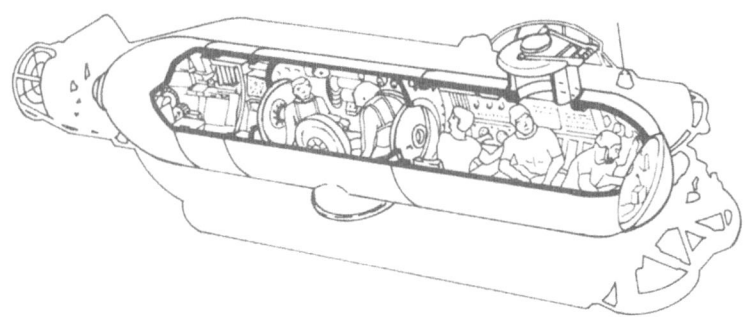

Figure 6.3 L5

ensuring that the vehicle can carry sufficient payload for support and rescue operations.

The satisfaction of mating requirements and the support of field inspection and maintenance tasks will require the incorporation of additional operational equipment. A vehicle with an operational capability less sophisticated than the Deep Sea Rescue Vehicle (D.S.R.V.) (Fig. 6.2) and slightly better than the present L5 or PC 1800 (British Oceanics) (Fig. 6.3), would generally appear suitable for the purpose, although a smaller submersible incorporating the salient features may be preferable because of surface support restrictions.

6.4.1 Pressure Hulls

The internal volume of the pressure hulls is dictated by the operational equipment to be fitted to perform the tasks, and personnel and materials to be carried. These parameters need to be established to allow optimization of the shape of the pressure hulls. Generally, for our depths of operation, pressure hulls should be tested at 1.5 times maximum operating depth and have an ultimate collapse depth of at least twice maximum operating depth.

Design of the pressure hulls should consider the influence of penetrations, mating systems, corrosion, collision and hydrodynamic forces as well as the net hydrostatic sea pressure. During dry transfer operations the stress distribution around the pressure hull will change. A collision may destroy the stress balance and cause catastrophic failure of the pressure hulls.

For our task requirements it is advisable to use a configuration with a minimum size control sphere which can be temporarily isolated by a hatch from an end-capped cylindrical pressure hull, incorporating the mating skirt. This arrangement would allow maximum volume for personnel and materials transport and isolate the control sphere from possible leaks at the mating surface during ingress/egress operations to the complex. The pressure hulls could be made of metal or reinforced fibre-glass, the latter providing payload advantages. If the use of a smaller submersible was considered advantageous the minimum volume control sphere could be retained and the other hull reduced in size.

6.4.2 Ballast System

The ballast system of a submersible is generally broken down into two categories; fixed ballast and variable ballast. They provide buoyancy and attitude control. Fixed ballast is composed of fixed flotation material and/or ballast weight. The variable ballast system normally consists of three separate systems, main ballast, auxiliary ballast and a trim system. Capacities of various ballast systems are governed by the vehicle's stability.

The main ballast system is usually used to achieve neutral buoyancy for diving by flooding tanks, or positive buoyancy by expelling seawater with high pressure air. Auxiliary or variable ballast can be used to compensate for the contraction of the vehicle under pressure and for changes in seawater density to maintain neutral buoyancy at all depths. The trim system, although considered part of the ballast system will not change the net buoyancy of the vehicle, but will alter the longitudinal axis in the vertical and, sometimes in more complex vehicles, in the lateral axis.

In a submersible used for dry transfer operations all ballast systems will be required. The trim system will be necessary to assist mating operations. The transport of personnel and materials will involve large buoyancy changes, so loads removed from the vehicle to the complex should be balanced as closely as possible by personnel and waste loads to be returned to the surface, to retain neutrally buoyant conditions without ballast adjustment. To take on ballast at depth to make up for reduced return loads, should be possible, but in the rescue role, special ballast will have to be carried during the descent phase to compensate for rescuees. Water ballast could be carried that is drained to the complex as a crew member moves into the submersible.

6.4.3 Propulsion and Manoeuvring System

The main propulsion system required for the support vehicle will be relatively standard as endurance requirements will be minimal for the ascent and descent phases of operation. Manoeuvring propulsion will however, be required to assist mating operations and subsea maintenance and inspection tasks. By providing vertical and transverse thrust these systems will allow a degree of control over the vehicle attitude and direction.

Forward and reverse propulsion is normally provided by a single screw propellor with a rudder mounted in the propellor wash. Simple dynamic control planes are mounted ahead of the rudder or on the bow for controlling pitch attitude. Shallow operating submersibles can use propulsion shafts that penetrate the personnel or materials pressure hull, but for deep operations the entire propulsion system is normally mounted outside the hull to minimize hull penetration.

Manoeuvring systems are usually an extension of the main propulsion system and may be auxiliary thrusters, multiple ducted thrusters or variable thrust vector devices. Many submersibles now incorporate a hover facility which integrates feedback from sensing devices with propulsion effectors (stern propellor and ducted thrusters) to maintain the submersible at a mid-water position.

A submersible for the task requirements will probably require a propulsion and manoeuvring system capable of providing 4-5 knots forward speed, a hovering capability and an ability to control the submersible in attitude and position against a 1 knot current. Propulsion power systems endurance will not be a critical element in the submersible design, because of the limited deployment period and the convenience of support systems.

6.4.4 Submersible Control

The control systems of the submersible should provide the pilots with complete control over the angular and linear movement of the vehicle. The pilot of the vehicle is mainly concerned with three basic vehicle manoeuvres; one, to cruise directly and as quickly as possible to the underwater complex from the surface or subsea base and return, two, to hover at the required attitude over the complex docking plate and move slowly into contact with it and three, to pilot the vehicle across the seabed, avoiding obstacles, to the work site location and hover while performing work tasks or deploying other systems.

Two regimes of control are therefore required, the cruise regime and the slow speed or hovering regime for mating, maintenance and deployment tasks. In the cruise mode conventional rudders or shrouds can be deflected to move to port or starboard and planes or shrouds deflected to change the ascent or descent path. In the hover mode or during slow speed manoeuvring, horizontal and vertical thrusters can be used to vary the position, and the trim ballast system activated to alter attitudes.

Many navigation aids and sensor devices will be available to assist submersible control operations, and, with proper design, these should minimize submersible control problems. The most sensitive area for control operations will be encountered during mating operations to the complex; these operations should be optimized to minimize the possibility of collision and to reduce impact loading between the submersible and the complex.

6.4.5 Mating

To perform mating operations and one-atmosphere transfers the submersible will require the fitting of special systems. The submersible will require a bottom hatch in one of the pressure hulls and a specially designed mating or transfer skirt attachment. The skirt is normally hemispherical and made of the same high strength material as the submersible pressure hull and is attached to the flange of a stub skirt surrounding the submersible bottom hatch. A vertical fin is sometimes fitted aft of the transfer skirt to reduce turbulent wake set up by the skirt assembly when the submersible is cruising. (This will probably not be necessary for our operations.) A shock mitigation ring may also be incorporated around the skirt to reduce impact loading on docking to the structure.

One method of docking is to actuate a hydraulically operated pump in the skirt as the submersible comes into close proximity with the docking plate. This begins to remove water from the skirt to sea. As the seal between the transfer skirt and the docking plate becomes effective, differential pressure causes the pump to stall and the overflow from the pump is redirected into the transfer ballast tanks fitted externally to the submersible hull. As positioning aids, magnetic anchors could be fitted to the complex or a special landing base with guide-rails installed. The submersible may also be fitted with skids that allow it to lower itself onto the docking plate when it is in the correct location. The manipulator arm of the submersible can be used to clear marine fouling and debris from the sealing services or engage a grapnel into the complex hatch bail for pull-in operations in strong currents.

Once the submersible is located correctly and the transfer skirt is dewatered, the lower hatch of the submersible can be opened, and any physical hold-downs necessary can be made. The complex access hatch will contain valve arrangements for testing pressures and atmospheres. On completion of such tests, the access hatch is opened and one-atmosphere transfers from the submersible to the complex can be performed. On completion of operations, hatches are closed, the transfer skirt is rewatered, and buoyancy is adjusted for submersible float-off.

Such docking procedures are at present standard operational practice on an intermittent basis. The implications of day-to-day operation of such systems may require investigation. The pressure differential between ambient sea pressure and the one-atmosphere dewatered transfer skirt provides a very effective retaining force for the seal, the effects of current drag forces or seismic disturbances on the seal should, however, not be ignored. Bottom mounted transponded navigation systems should provide location accuracies to the docking plate of the order of ±1 m, a clear path of approach for the submersible should, however, be maintained.

One major disadvantage of the incorporation of the transfer skirt on the submersible is that at launch, air can become entrapped in the skirt upsetting the buoyancy adjustment. The entrapment of air in the skirt at depth during separation from the complex, will be even more critical, as this air will expand during ascent causing a rapid change in the buoyancy. Correct operational procedures should be implemented to avoid such occurrences.

6.4.6 Power Requirements

Electrical power for propulsion, hydraulic circuits, ballast control and instrumentation systems will probably be provided by a battery storage system. For such deep-water operations the system should be oil compensated for pressure and stored externally to the hull, to save the internal space available for logistic items. An emergency power supply should, however, be stored internally for control purposes should the external systems fail.

Hydraulic energy will probably be provided by electrically driven pumps and the pressure raised converted to mechanical energy for operating the mating pump, manipulator arm, ballast system and control systems. The dewatering pump may represent a significant power load for normal submersible operations, however, due to the limited need for propulsion power, this requirement should not be a major restriction to the system design.

6.4.7 Sensors

Sensors will be required to monitor the internal conditions and operation of the vehicle and sense external conditions. Internal sensors will monitor tachometers, thermostats, flowmeters, transducers, power consumption and smoke and leak detectors etc. The submersible is likely to be fitted with a plexi-glass dome for efficient viewing by pilots, however external television cameras will be required to extend the field of view and monitor the mating operations. Mercury vapour and quartz iodine lights will be fitted to provide illumination for viewing in the darkness of the depths.

Sonar devices, as well as those used as navigational aids, will be used to detect obstacles in the path of the submersible and for locating the complex on the seabed. Vertical obstacle avoidance sonar, sweeping vertically, will give an indication of the height from a reference plane of an object in the submersible's path, and horizontal obstacle avoidance sonar will indicate the range of objects in the horizontal sweep plane. Short range sonar will indicate altitude and range on approach to the docking plate on the complex.

Meter dials, graphic displays, sonar displays, television monitors and indicator lights will display the parameters in the control sphere for the pilots to monitor and assimilate for control purposes.

6.4.8 Navigation

Submersible navigation is usually relative to a fixed bottom position and in this mode it is more efficient to navigate in terms of range and bearing to the complex. Navigation equipment may vary in complexity from an inertial navigator to a simple compass, depth gauge and current indicator.

The position of the submersible in the water column may be obtained from a pressure gauge (depth meter) or depth altitude fathometer. Transponder acoustic signals can be used to determine range and bearing. A tracking transponder aboard the submersible can be used to emit a sonar signal upon interrogation by a surface support vessel, this provides an accurate means of tracking the submersible during diving operations. Alternatively, a transponder interrogation sonar aboard the submersible can interrogate and receive signals from a transponder placed on the underwater complex or the surface support vessel.

Other sonar devices are also available to assist navigation. Doppler sonar can be used to measure vehicle velocity fore/aft and port/starboard in knots and vertical speed in ft/min, a sound velocimeter may also be needed to provide a continuous calibration of the speed of sound in water to account for changes in seawater temperature and pressure. Obstacle avoidance sonar has been discussed previously.

Sonar transponder navigation systems will be adequate for the

location of the complex and surface vessels, although thermoclines in the water depths of operation may cause complications.
For the descent/ascent phases, which are repetitive in nature, the
introduction of computerized automatic control of these operations may improve descent/ascent transit times and enhance safety.
Guidance to the docking plate will probably be achieved to within
±1 m, short range sonar devices will be necessary to achieve the
fine tolerances for mating operations. Transponders providing
sonar ranges and bearings will be necessary for bottom navigation
tasks, and obstacle avoidance sonar will be critical for field maintenance and inspection tasks.

6.4.9 Communications

The ability to communicate with surface support vessels, other
submersibles and the underwater complex will be critical to the
safe and efficient operation of the submersible. Communication
devices will probably provide voice or continuous wave keyed
contact. Submerged, the submersible can use the underwater
telephone in either mode to communicate to surface vessels, the
complex or other submersibles. A directional listening hydro-
phone aboard the submersible will permit the vehicle to home in
on a source of sound. Surfaced, UHF radio will provide a two-way
line of sight communication to the support vessel, and, if the
radio is switched to transmitting continuously, this will provide
a signal for the surface vessel to home in on. An intercom should
also be provided between the two pressure hulls of the submersible
for communication when the interconnecting hatch is closed.

6.4.10 Life Support

Life support systems will generally be as normal, i.e. removal of
carbon dioxide and contaminants and make up with oxygen. Life
support systems' capacity will have to be sized to account for the
additional requirements for personnel carriage in the normal
transfer mode and in the rescue role. Emergency life support
systems for the submersible should also take these factors into
account.

6.4.11 Support Systems

Surface and subsea support vessels for the operation of the submersible are dealt with in a later chapter in this book. Service
facilities required for the support of the submersible are, however, mentioned here.

Support requirements from the vessel include, battery charging,
life support replenishment and systems checkout as well as the
normal operations of loading and unloading personnel and materials.
Special interface units for servicing power and ballast systems will
be required to provide efficient support of the vehicle and reasonable turn-round times between supply missions.

6.5 BELLS

The contents of this chapter have generally concentrated on the
operational aspects of the use of a submersible as the underwater
support vehicle for the complex. A suitably designed diving bell
could also provide a suitable vehicle for the support tasks. A
unit deployed from a surface support vessel, preferably fitted with
manoeuvring thrusters, could mate with the docking platform on
the complex in a similar manner to the submersible and perform
one-atmosphere transfers of personnel and materials. A bell could
also provide a rescue facility and perform limited subsea maintenance and inspection tasks.

The use of such a concept may, however, involve many un-
desirable complications. The pendulum effect induced in the
suspended bell by surface movement of the support vessel may be
excessive in rough weather and over the deployment depths
envisaged. Subsea currents will tend to displace the bell from its
deployment path due to the imposition of hydrodynamic drag
forces on the unit. The surface vessel would also need to be
anchored directly over the complex, endangering the complex if
the anchors should drag.

The operational concept of bell support is relatively simplistic
and therefore is not considered here in detail, but is followed up in
a later chapter on support systems for subsea operations. In the
context of the overall task requirements for the vehicle, the use
of a submersible would appear most appropriate. The bell system,
however, may have advantages for the transfer of large loads to

the complex through oversize hatches, where deployment operations are not critical and submersible payload capabilities are limiting.

6.6 AUTONOMOUS SUBMARINE

An alternative concept of submersible support vehicle is that of the autonomous submarine. The vehicle would be capable of transporting men and materials from a port base to the underwater complex by submerged transit. The concept has been conceived to avoid the launch and recovery limitations of surface support systems. The submarine would dock with the complex in a similar manner to the use of normal submersibles, but under stricter control due to the large mass of the vehicle, and perform one-atmosphere transfer operations of personnel and equipment to the structure. It would load off-going crew and waste materials and return to the home port again submerged. The vehicle could also perform the rescue task, and to a more limited extent, inspection and deployment tasks.

Such a concept of support vehicle is presently under detailed study by Slingsby Engineering for the Deep Sea Production System Consortium and, as such, the reader is referred to Reference 23 for further information. The cost of such a system may, however, be extremely expensive.

INGRESS/EGRESS SYSTEM FOR COMPLEX

7.1 INTRODUCTION

The support of the manned underwater structure by an underwater vehicle will involve careful design of the interface between the complex and the support vehicle. Access to the structure from the underwater vehicle must be designed such that the integrity of the structure is retained during transfer operations of crew and materials through the interface. Potential hazardous situations must be avoided and suitable safety design features incorporated.

Support operations through the interface should be achieved as efficiently and as quickly as is feasible. The ingress/egress system design should also take into account the need to rescue personnel in an emergency situation and the possibility of having to transfer unconscious personnel through the interface into a rescue vehicle.

Logistic support loads of personnel and materials need to be evaluated for their impact on system design and the environmental parameters at the site location incorporated into safety design studies. The requirement for the provision of a safe haven in the underwater complex should also be correlated with access facilities.

7.2 BASIC PHILOSOPHY OF THE INGRESS/EGRESS SYSTEM

Elements considered essential in the design of a suitable ingress/ egress system for the underwater complex are:

1. Safety of crew during transfer operations is of major concern.
2. Most potential hazards will exist from the time the vehicle contacts the mating surface on the station until departure.
3. Underwater vehicle should be mated to the structure access for the minimum feasible time.
4. Transfer operations from the underwater vehicle to the structure and vice versa should be undertaken in the lock-in/ lock-out mode.
5. Lock-in/lock-out cycles should be minimized and be, preferably, a single cycle.
6. Lock-in/lock-out operations should be performed using a dedicated access module, independent of the main complex structure.
7. Hatch dimensions should be standardized, hatches should be operable from either side and close under sea pressure.
8. Access chamber should if possible, serve as a safe haven and as a one-atmosphere buoyant escape capsule during emergencies in the complex.
9. All systems should perform efficiently and operate in the fail-safe mode under expected environmental conditions and in emergencies.

7.3 INGRESS/EGRESS SYSTEM CONCEPT

The system has to provide the means of transferring on-going crew and fresh supplies to the underwater complex from an underwater vehicle and returning off-going crew and waste materials to the surface. The system should also allow the rescue of personnel from the structure and/or escape to the surface. The major constraints to such a system design are:

1. Configuration of the underwater structure complex.
2. Environmental conditions to which the system will be exposed.
3. Logistic support requirements, i.e. crew, materials, spares etc.

7.3.1 Configuration of Structure

The underwater complex is envisaged as being a collection of horizontal, hemispherical end-capped cylindrical modules. The design of the ingress/egress system should be compatible with such an arrangement of modules. If the ingress/egress system can also serve as a safe haven and a one-atmosphere escape capsule, this will negate the need for a safe haven to be incorporated in the complex structure or the attachment of a one-atmosphere escape capsule.

7.3.2 Environmental Conditions

The environmental conditions of operation have been stated in Table 3.2. At the maximum operating depths of the complex severe restrictions will be placed on pressure hull configurations because of the strength limitations of materials and the lack of knowledge of stress patterns in complex geometrical shapes. Spherical and end-capped cylinders should, therefore, only be considered for the addition of modules for the ingress/egress system. Currents and seismic forces can affect the hatch loading of the underwater vehicle, visibility may be restricted by turbidity and areas exhibiting turbidity currents should be avoided. Corrosion and marine fouling may also affect the sealing surfaces.

7.3.3 Logistic Support Requirements

The details of logistic support requirements are dealt with in Chapter 5.

As a philosophy of liberal design in life support systems is recommended to reduce the costs of support operations, resupply requirements are generally restricted to the movement of crew, supply of food and recreational materials for the crew and replacement of spares, lubricating oils and filters. Disposal of waste from the structure will be a direct function of the overboard dumping that is acceptable if it is assumed that sanitary wastes are disposed to sea.

Logistic loads have been estimated as approx. 5000 kg based on resupply for thirty men over a fourteen day period. It is envisaged that the underwater vehicle with a dry payload capability of between 750 and 1000 kg and with an available transfer volume of

4 m³ will be used to supply the structure using three dives per week.

7.4 ACCESS CHAMBER

A pressure chamber, independent of the main complex structure, should be provided to act as an intermediate transfer structure between the complex and the underwater support vehicle. The access chamber should have a hull capable of withstanding the hydrostatic pressure at the operating depth of the complex and internal volume at one-atmosphere pressure. The access chamber will also incorporate the docking and mating facility for the under-water vehicle. The docking of the vehicle to the complex itself is possible, but it is not recommended because of safety considerations. The use of an access chamber provides these additional safety features:

1. Allows lock-in/lock-out of personnel and materials to and from the underwater complex.
2. Provides a suitable volume for storage of materials and transfer of personnel prior to movement to the complex or the underwater vehicle.
3. Minimizes the time the underwater vehicle needs to be mated to the complex.
4. May provide two hatch entrances to the structure.
5. By suitable design could provide a time delay to allow crew to escape to the complex if a leak occurs at the mating surface.

Various access chamber configurations that are feasible are shown in Figs 7.1, 7.2, 7.3 and 7.4. The concept as depicted in Fig. 7.3 is preferred as it would provide maximum volume for transfer operations and the possibility of using the chamber as a safe haven and/or buoyant escape capsule. Figure 7.6 shows a conceivable design of such an access chamber based on the dimensions of a complex for 500 m operational depth. Figure 7.5 depicts a more simplistic concept for the minimal logistic loads envisaged in our studies.

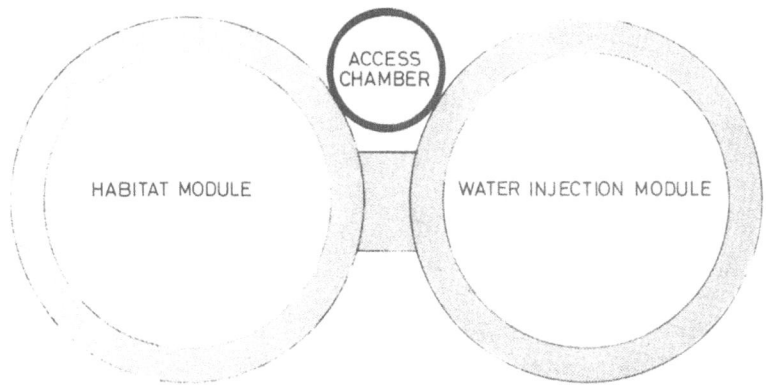

Figure 7.1 Ingress/egress system — single sphere access

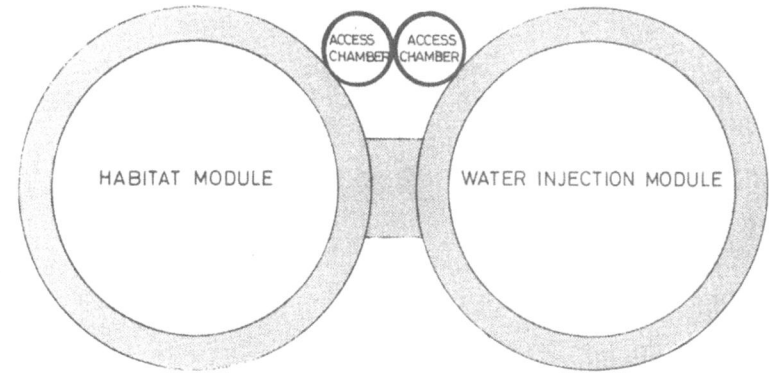

Figure 7.2 Ingress/egress system — adjoining twin spheres access

7.4.1 Access Chamber as Safe Haven and/or Escape Capsule

Previous studies of manned underwater structures rescue and escape systems,[1] have indicated that a safe haven should be provided in the complex and a one-atmosphere buoyant ascent capsule should be incorporated in the system design to provide a secondary means of rescue of the crew if underwater vehicle operations are not possible. An access chamber that could also provide a safe haven facility and escape system would appear desirable (Fig. 7.6).

To incorporate such concepts the access chamber would require

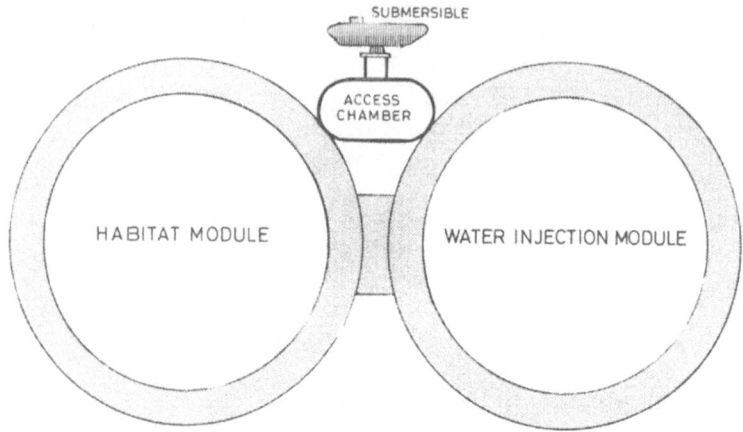

Figure 7.3 Ingress/egress system — hemispherical end-capped cylinder access

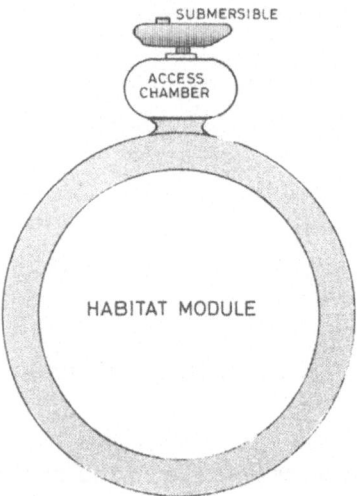

Figure 7.4 Ingress/egress system — simple access

specific design modifications. To perform as a safe haven, emergency life support requirements would need to be satisfied. As a one-atmosphere escape capsule, as well as additional life support systems, the capsule should have suitable surface sea-keeping characteristics and to achieve pressure tight storage of the chamber, the habitat and water injection modules would require special

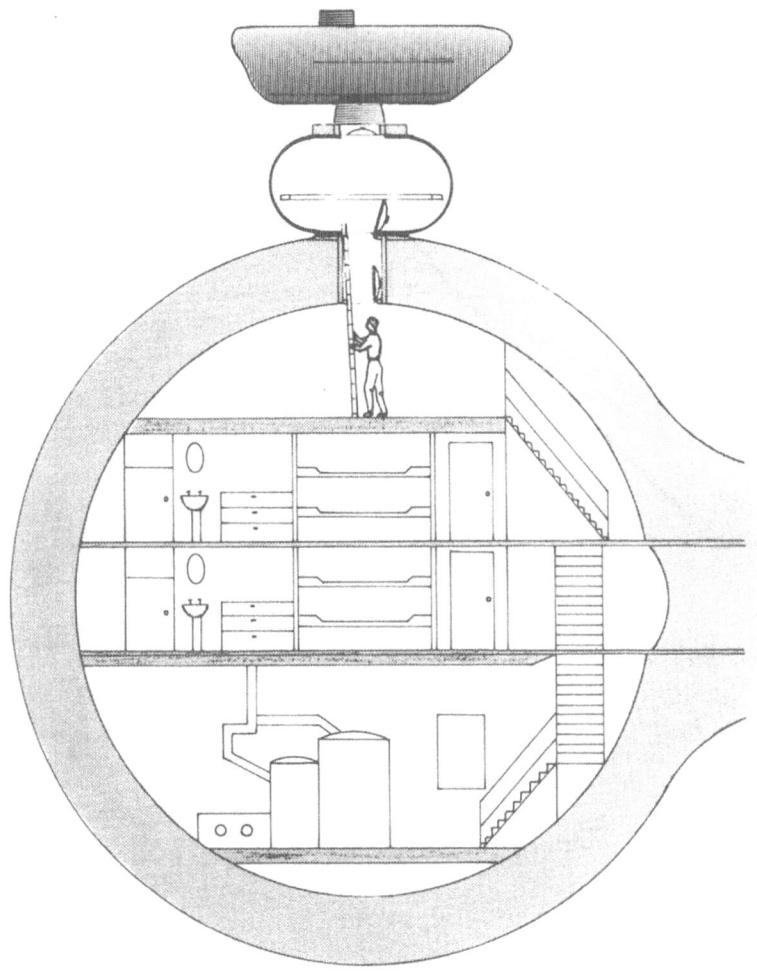

Figure 7.5 Simplified access concept

hatch coaming rings and sealings. The space between the hatches of each passage would be drained into the structure for checking purposes and the seal maintained by hydrostatic pressure.

In use as an escape capsule, once the crew have entered the access chamber, the chamber hatch and module hatches are closed and the intermediate space flooded, releasing the pressure and enabling the slipping of the positively buoyant sphere. Suitable safety harnesses and seating arrangements would be necessary to

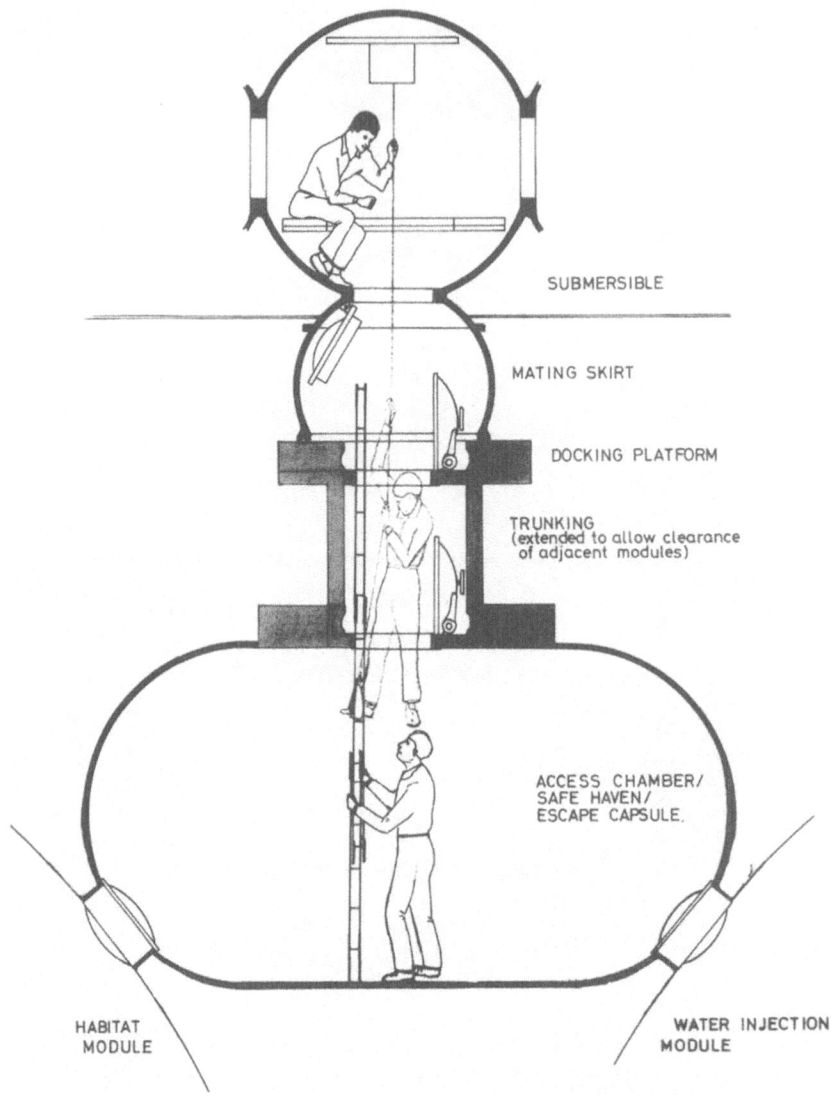

SUBMERSIBLE

MATING SKIRT

DOCKING PLATFORM

TRUNKING
(extended to allow clearance
of adjacent modules)

ACCESS CHAMBER/
SAFE HAVEN/
ESCAPE CAPSULE.

HABITAT
MODULE

WATER INJECTION
MODULE

Figure 7.6 Access chamber as safe haven and/or escape capsule

protect the crew during the ascent phase and in sea motion on the surface.

The incorporation of emergency life support systems and other ancillary requirements, for the chamber to act in the role of a safe haven and an escape capsule, would utilize minimum space in the chamber and should not restrict transfer operations. The major design consideration would be the safety design of the releasable hatch connections as these may introduce potential hazards similar to those at the mating interface. These seals however, would normally be permanently connected and only disturbed in an emergency. Suitable engineering design should satisfy safety considerations.

The major advantages of such a combined design concept are that valuable complex space would not need to be dedicated to a safe haven facility and separate escape capsules would not need to be fitted externally. The underwater vehicle docking plate located on top of the access chamber would be ideally situated for crew rescue from the integral safe haven. If vehicle operations were not possible for rescue, the crew could release the access chamber/ escape capsule and ascend to the surface in safety. An access chamber/escape capsule should be made available at each end of the module to provide total coverage of the habitat.

7.4.2 Access Chamber Services: Transfer Operations

During resupply operations the access chamber, except for brief periods, will be open *either* to the underwater vehicle or the habitat/water injection modules via 25 in diameter hatches. In the purely transfer role, life support requirements will be circulated from the vehicle or the modules, after an initial ventilation period. Power requirements will be limited to lighting and any electro-mechanical equipment used for assistance with loading operations.

Internal communications to the habitat will be provided by cable links and waterproof plug connectors to the station intercom system. External communications to the submersible vehicle and surface support vessels would be by underwater telephone.

7.5 RESUPPLY OPERATIONS

The major operational areas involved in the supply of crew and materials to the structure are:

1. Transit of materials and personnel from the support base to the surface at the site location.
2. Deployment of the underwater vehicle through the water

column to the access chamber and recovery on return to the surface.

3. Transfer of materials and personnel from the underwater vehicle to the structure and removal of waste materials and off-going crew from the structure.

Items 1 and 2 are treated elsewhere in this book. Transfer operations are treated now in detail with reference to ingress/egress system design considerations.

7.5.1 Transfer Operations

The access chamber acts fundamentally as a buffer storage volume during the transfer of personnel and materials to and from the structure. Ideally, the chamber should be capable of storing materials ready for removal and the incoming vehicle payload. As vehicle payloads will be relatively restricted, a suitable design of access chamber should have sufficient volume to store the payload. If the access chamber volume was limited, multiple transfer operations from the vehicle to the complex would be necessary. The use of smaller vehicles to improve surface handling operations will involve reduced vehicle payloads that may only be a segment of the transfer load requirements. Multiple resupply cycles by the vehicle may, therefore, be required.

The transfer of materials and personnel to and from the structure should be performed as a lock-in/lock-out operation. Habitat/water injection hatches will be closed when the access hatch to the vehicle is open, and the access hatch should be closed before the habitat or water injection hatches are opened. Hatches to the complex should be located as far up the side of the access chamber as is possible, to allow time for escape through them if a leak should develop while the access hatch is open.

7.5.2 Sequence of Transfer Operations

The generalized sequence of events for transfer operations is shown in Fig. 7.7, which gives a simplified flowchart of operations. It is assumed that the chamber can accept the full payload from the underwater vehicle, but does not allow storage of out-going

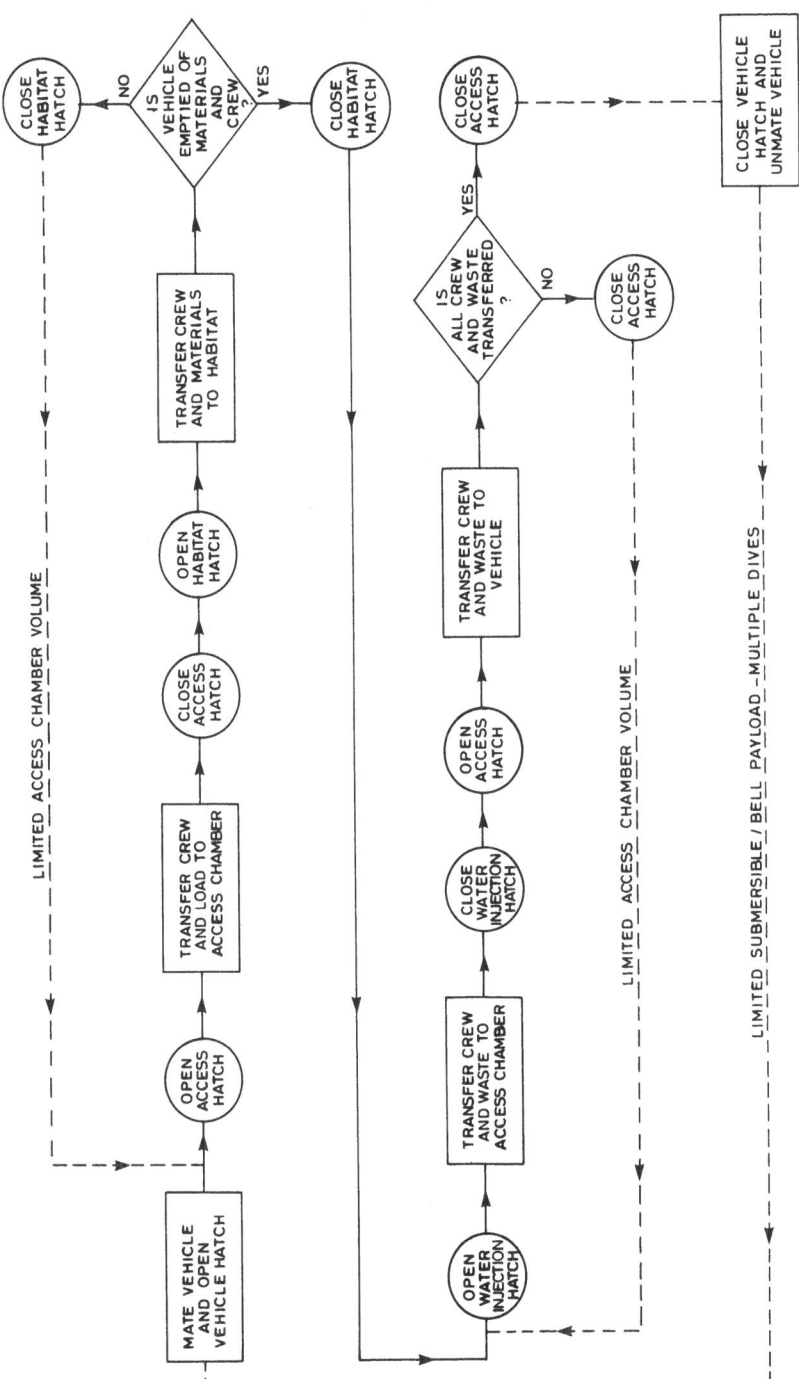

Figure 7.7 Flowchart of transfer operations

materials prior to removal. Estimates indicate that a 2–3 hour period would be required for a single resupply cycle.

1. Vehicle lands on mating platform, adjusts ballast, dewaters skirt and tests pressure and atmosphere.
2. Vehicle crew member opens vehicle hatch and equalizes pressure between access chamber and vehicle and tests chamber atmosphere.
3. First replacement crew member opens access hatch and enters access chamber (by hoist), performs inspection and establishes internal communications and erects access ladder.
4. Portion of on-going crew enters access chamber and remainder of crew pass materials to crew in access chamber.
5. Remainder of on-going crew in vehicle move into access chamber, remove ladder and close access hatch.
6. Habitat hatch is opened, supplies and materials are transferred to habitat and moved to storage by on-going crew.
7. Rest of on-going crew moves into habitat and habitat access hatch is closed.
8. Off-going crew in water injection module then open access chamber hatch and portion of crew enters chamber.
9. Waste material and other things to go to the surface, which are stored in water injection module, are loaded by off-going crew into the access chamber.
10. Remainder of off-going crew moves into access chamber and water injection module access hatch is closed.
11. Access hatch to vehicle is opened after testing, ladder is erected and a portion of crew enters vehicle. Waste materials are loaded aboard the vehicle.
12. Off-going crew left in the access chamber moves into the vehicle, the last man removing ladder before being hoisted into vehicle. The hatch to the access chamber is closed.
13. Vehicle crew close vehicle hatch, skirt is rewatered and ballast adjusted and the vehicle ascends to the surface.

The feedback loops in Fig. 7.7 indicate those sequences that have to be repeated if the vehicle payload is smaller than the transfer load, or the access chamber volume is limited. In our concept it is unlikely that the access chamber volume will be restrictive. The choice is, whether large payload vehicles with the attendant handling problems, high surface support costs and relatively long locked-on periods for single cycle load transfer is better than reduced surface operational problems and costs with smaller vehicles, making multiple dives with smaller payloads and spending less time locked-on to the access chamber.

The concept proposed of combining access chamber/safe haven/ escape capsule should provide sufficient volume of storage to allow a single cycle transfer of materials and personnel. If smaller vehicles are used, multiple deliveries to the access chamber could be stored prior to movement into the habitat. The concept would also provide double access to the complex, which could be used to improve the speed of material transfer.

Fundamentally, the ingress/egress system should provide lock-in/lock-out of men and materials to the complex, whilst retaining a closed hatch at all times between the mating skirt and the habitat and water injection modules, to confine flooding in the event of a leak. To minimize the risk to the complex the support vehicle should be mated for the shortest time possible by optimizing the efficiency of actual operations and balancing the number of transfer cycles against docked time.

7.5.3 Transfer of Injured Personnel from Complex

The safe transfer of injured personnel from the structure will depend on the degree of locomotion the injured crew member is capable of. Generally, most injuries that could be expected will still allow the crew member to use the ingress/egress system, however, those incapable of movement will need special assistance. The injuries that can produce lack of self-locomotion are:

1. Unconsciousness.
2. More than one limb broken.
3. A broken back.
4. Burns over most of the body.
5. Shock.
6. Paralysis.

A powered hoist or block and tackle should be provided in the underwater vehicle to lift the casualty vertically up into the vehicle from the access chamber. A modified parachute harness could be used as a sling to provide support at the crotch, around the chest and at the back of the casualty. A block and tackle would have to be operated by other members of the crew in the access chamber if a powered hoist was not available. Portable ladders available for ingress and egress by hatches could be used as a stretcher to move the casualty from the habitat/water injection module into the access chamber if no other means was available.

7.5.4 Hatches

All hatches in the ingress/egress system should be arranged so as to seal with sea pressure. The thickness of hatches will generally be thicker than the section of the equivalent metal pressure hull to allow for any lack of sphericality in hatch shape.

The ingress/egress hatch between the underwater vehicle and the access chamber will probably be a spring loaded seal hatch incorporated in a stiffened foundation at the top of the access chamber, under the mating platform. Hatches will be similar in design to submarine hatches, but heavier to provide for the deeper depths of operation. Double operating gear should be provided so that hatches can be opened from either side, with high pressure packing placed around the operating gear, which is only used at one-atmosphere so no binding should occur. Penetrator valves will be provided in hatches to permit pressure equalizing and atmosphere sampling.

The hatches leading to the habitat and water injection modules will be of similar design to the access hatch as they need to be operated from both sides and must be capable of withstanding ambient sea pressure. These hatches will be set in heavy ring forgings located at the chamber/module intersections. Detailed stress analysis of this area will be required during design. If the access chamber is to act also as an escape capsule, a special double hatch will need to be developed at the intersections.

The use of standard naval submarine hatches will provide hatch openings of 0.635 m (25 in). Hatch collars of thick rubber or semi-rigid fibre-glass will need to be provided to protect machined sealing faces during loading operations.

7.5.5 Dimension Restrictions

The maximum size of packages and equipment that may be transferred by the ingress/egress system will be determined by the
minimum size hatch opening in the station or support vehicle. If
standard naval 0.635 m diameter hatches are adopted, square
packages not exceeding 0.38×0.38 m (15×15 in) should be
used to ease handling through the hatches. Cylindrical packages up
to a diameter of 0.42 m (16.5 in) could also be used, but refrigera-
tion storage systems are normally designed for square packaging.
Length restrictions are more difficult to predict, but will generally
be determined by the dimensions of the vehicle used for support
operations and its capability to carry such items.

The weight of items, if they are to be handled by one man,
should be restricted to less than 16 kg (35 lb). The use of hoists or
pulley blocks may allow heavier items to be lowered vertically
from the support vehicle into the access chamber and then man-
handled into the habitat or water injection module on a rolling
loading rack.

Waste materials, samples etc., that are to be removed from the
complex will also be restricted by the same parameters. Containers
used to supply materials should be capable of being used to store
waste materials before removal. The supply and removal of larger
items will necessitate expensive development of larger hatch sizes
or specialized underwater docking systems for subsea modules.

Detailed consideration of the storage of materials and seating of
personnel in the submersible and access chamber is required.
Spherical pressure hulls will not be the most effective shape for
storage. The use of revolving loading platforms in spheres will
increase packing density to the order of 70%. The proposed use
of an end-capped cylindrical structure for the access chamber, will
maximize the volume available for transfer purposes, safe haven
and escape capsule requirements.

7.5.6 Mating to Access Chamber

It is envisaged that the access chamber will be fitted with a specially
designed mating platform similar to those used for docking underwater vehicles to fleet submarines (D.S.R.V.). The special mating
skirt fitted to the submersible vehicle seats on the machined

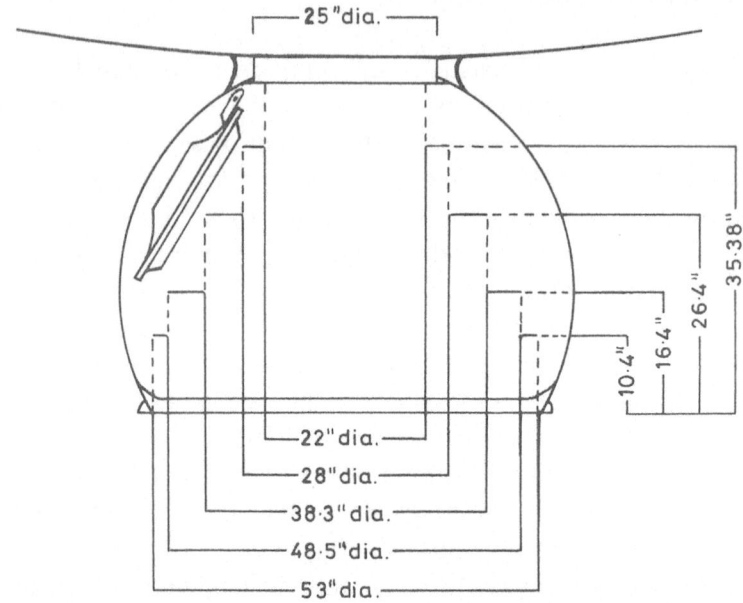

Figure 7.8 D.S.R.V. skirt dimensions

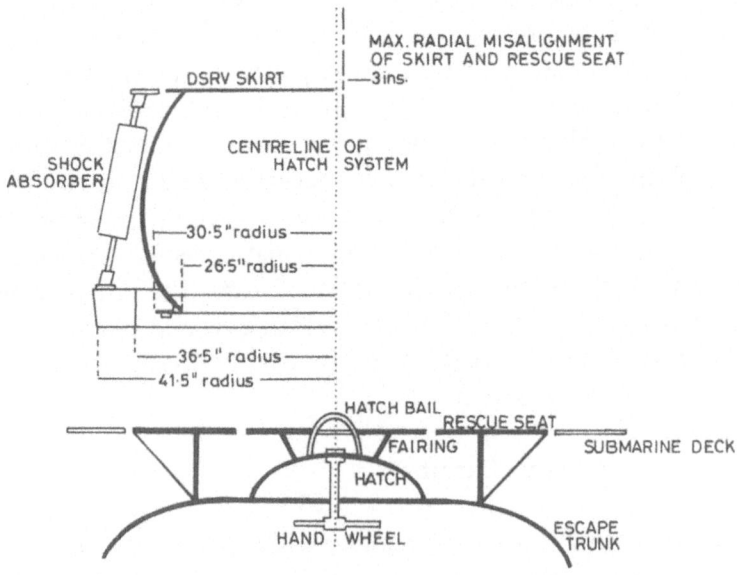

Figure 7.9 Details of typical submarine rescue seat

face of a low profile docking platform surrounding the access hatch. On mating, the submersible pumps out the entrapped water into a tank in the vehicle superstructure and thereby effects a pressure differential, ambient pressure holding the skirt and submersible on the access chamber. At this stage pressures are tested and the submersible hatch and access chamber hatches opened.

The access chamber hatch has to open up into the skirt of the submersible, so its dimensions are critical (Fig. 7.8). For the D.S.R.V. the docking seat must have a minimum outer diameter of 1.65 m (65 in) and a maximum inner diameter of 1.13 m (44.5 in). Additionally, the area beyond the docking seat must be in the same plane as the rescue seat and clear of obstructions and projections out to a diameter of 2.26 m (89 in) to accommodate the shock mitigation ring on the submersible (Fig. 7.9).

The skirt mating flange contains a rubber gasket and will seal over irregularities up to 0.004 m (0.15 in) at operational depths and under loads impressed by the submersible. In heavy current conditions a grapnel and hauldown winch can be used to assist mating if a suitable bail is available on the access hatch.

7.6 POTENTIAL HAZARDS

Most potential hazards to personnel exist from the time the support vehicle contacts the mating seat until the vehicle departs, and involve retaining the integrity of the seal between the submersible skirt and the docking plate.

7.6.1 Impact Collision

Excessive impact between the vehicle and the station can damage the mating skirt and the docking seat, and, in extreme circumstances, may disturb stress patterns in the vehicle hull or access chamber causing a failure. Every attempt should be made, therefore, to effect an easy landing, and shock mitigators may need to be fitted to the underwater vehicle to absorb impact energy on mating. The mating interface design should incorporate safety factors to account for the expected impact velocities and their associated loadings.

7.6.2 Currents

Strong currents will impose hydrodynamic drag forces on the vehicle while seated on the docking plate; overturning movements may be created that may break the seal. Operational areas that exhibit turbidity currents should be avoided.

7.6.3 Seal Failure

On dewatering the vehicle skirt, water vapour will be created in the skirt if the seal is successful due to the depressurization. Repeated failures to achieve these conditions normally indicate that an obstruction has occurred on the seat and this may be due to marine fouling or the entrapment of debris.

A failure of the seal once the seal is made will cause major problems. A small leak admitting a tiny stream of water at 2000 m depth (2940 psi) will impair visibility in the access sphere, as the water will vaporize immediately on expansion in the chamber. Hydraulic jet action will also quickly erode the gasket admitting even more water and requiring rapid escape by personnel into adjoining modules. The interconnecting hatches therefore, should be placed as high as possible in the access chamber to provide a time delay before water reaches the escape route, i.e. long enough for men to open the hatch, exit and close the hatch before water inhibits hatch operation.

7.6.4 Hatch Protection

The external hatches and the docking plate should be protected against marine fouling and obstructions. Internal hatches, when open and in use, should have their sealing surfaces fitted with protection collars. No obstructions should be placed in the way of hatch operations, i.e. electric cables, air ducting, in case the hatches need to be closed quickly in an emergency.

7.6.5 Warning Devices

As the access chamber is normally unattended except for transfer operations or emergencies, leak detection devices should be placed at the bottom of the chamber to detect the presence of water.

8

SURFACE AND SUBSEA
SUPPORT SYSTEMS

8.1 INTRODUCTION

Surface and subsea support systems will be required to service the manned underwater complex. Vessels will be needed to transport the products from the production complex and supply personnel and materials to the site location from a shore base. A loading buoy, tower or permanently moored tanker may also be used to act as a support platform to the complex, providing hotel accommodation, helicopter landing pad and a safety watch for the complex.

The characteristics of the vessel and/or platforms used for these task requirements will be determined by the operational constraints of the system in use and the prevailing environmental parameters at the site location. The distance of the site location from the shore base will establish the support frequency that can be expected depending on the support vessel's cruise speed. The deployment and recovery of underwater vehicles from and to a surface support vessel will be a critical operational area, primarily determined by the vehicle size and weight.

The surface and subsea support systems will represent a major capital investment in terms of field development. Considerations should therefore be given to the multi-functional use of such vessels to match the task requirements. Various support system

97

concepts are now considered in relation to sub-systems investigated earlier in this book.

8.2 UNDERWATER VEHICLE LAUNCH AND RECOVERY

The launch and recovery of an underwater vehicle to provide logistic support to the complex is fundamental to the system design, and involves a critical operation area. It is not possible at present to specify what surface support craft will be available at the site location. The choice will however probably lie with the following options:

1. Conventional submersible support vessel with an A-frame arrangement.
2. Permanently moored tanker with a moonpool facility.
3. Semi-submersible platform with a moonpool/cursor facility.
4. SWATH vessel with a moonpool/cursor launch.

The problems experienced in the use of such systems for the launch and recovery of the underwater vehicle are essentially determined by the size and weight of the underwater vehicle in combination with the prevailing environmental conditions and support vessel characteristics. Only by considering these parameters for a specific system configuration can a proper evaluation be made and the practicality of the system assessed. It is possible, however, to make some general points based on past experience.

8.2.1 Submersibles

The investigation of logistic support loads in Chapter 4 indicated that a submersible vehicle with a dry payload capacity between 750 and 1000 kg is required to resupply the complex on the basis of three dives per week. Physically this would require a submersible vehicle slightly larger than the British Oceanics L5 or PC1800 submersible and somewhat smaller than Taurus. A purpose designed submersible for the task requirements may, however, allow an overall reduction in size and weight of the vehicle, although it should be remembered that the additional equipment required for dry transfer operations may account for a portion of the dry payload capacity. A more frequent supply cycle could also be initiated to allow the use of a smaller submersible if launch and recovery conditions are limiting.

Present capabilities for submersible launch and recovery with conventional submersible support vessels are generally limited to sea state 6, wave height 4 m (5.5 m max.) for submersibles such as L5 and smaller, although in swell or mixed seas, operations may be limited to lower sea states. Larger submersibles, such as Taurus, can be handled by support vessels like British Voyager from an A-frame, but, because of handling difficulties with such a large mass, are restricted to sea state 5.

Reference 24 provides a detailed study of the operational problems associated with the use of manned submersibles in the support of a subsea manifold centre. Table 8.1 from this study gives an indication of the weather downtime that can be expected in operations in the North Sea (Brent Field: December 1975 to December 1978). This shows the number of 24 hour periods in each month where the wave height was less than 4 m (sea state 5).

TABLE 8.1 Weather downtime: submersible operations. British Oceanics Ltd., Brent Field Dec. 75-Dec. 78. Maximum wind speed 22-27 knots (Beaufort 6), Max. wave height 4 m (Sea State 5).

Month	Estimated number of operating periods	Number of calms significant wave height 4 m
January	6	9
February	12	17
March	12	18
April	17	25
May	20	30
June	20	29
July	20	30
August	20	30
September	15	22
October	15	22
November	10	15
December	12	17

The figures have been reduced by 30% to give a more realistic estimate of the weather downtime, due to other adverse conditions such as high winds, poor visibility and mixed and confused seas that can affect operations. The figures obviously only give a general indication of operation capabilities, as they will be a function of the submersible, launch and recovery system and support vessel characteristics.

January is obviously the most restrictive month for operations, but with a requirement for three dives a week of average duration three hours, it should still be possible to complete the necessary resupply cycles in this month. Liberal design of the complex storage could also be used to alleviate weather downtime problems as discussed in Chapter 5.

The data is obviously specific to North Sea conditions. The envisaged site location for the production complex is likely to be in more exposed waters over the continental slopes and at a distance from shore. It may therefore be necessary to consider the use of more stable surface platforms, such as semi-submersible vessels with a moonpool or cursor launch and recovery system to increase the number of acceptable operating periods under these conditions.

The presently available systems for submersible support would appear suitable for the resupply requirement with the use of the size of submersible considered. The use of larger submersibles (Taurus) and operations in more hostile environments than the North Sea, may require special stable platforms and launch and recovery arrangements.

8.2.2 Bells

Several diving bells or capsules are available that can perform one-atmosphere transfers of men and materials to an underwater structure. The bell is normally deployed from a surface support vessel via an A-frame over the stern or by an over-the-side launch from a deck crane. The bell is connected back to the vessel with an umbilical and backhaul cable.

Considerable experience has been gained by CanOcean Resources in the use of one-atmosphere service capsules for field operations on subsea completions and manifold systems. The capsule is deployed from a surface support vessel from an A-frame and

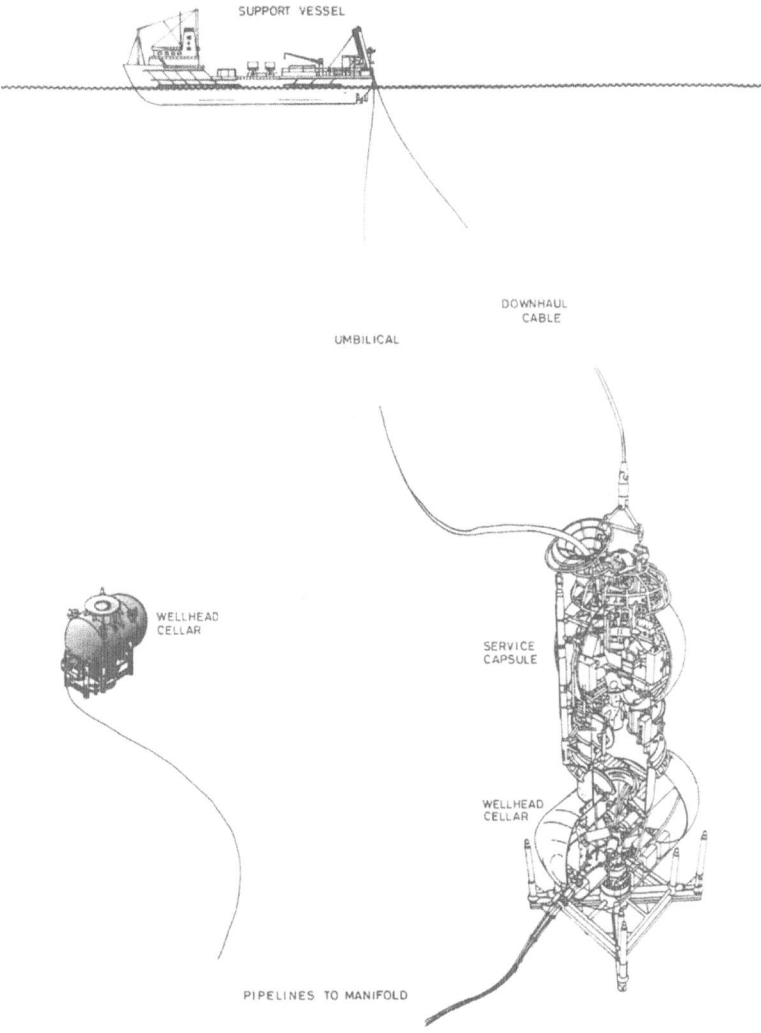

SUPPORT VESSEL

DOWNHAUL
CABLE

UMBILICAL

WELLHEAD
CELLAR

SERVICE
CAPSULE

WELLHEAD
CELLAR

PIPELINES TO MANIFOLD

Figure 8.1 Wellhead cellar (CanOcean) and support ship

winched down with the aid of a downhaul cable to mate with the
underwater structure (Fig. 8.1). The capsule can carry a maximum
crew of five and a payload of the order of 1000 kg. In open sea
the capsule can be launched in up to sea state 4 and recovered in
sea state 6.

Comex diving have also developed a series of one-atmosphere
transfer capsules (Fig. 8.2); the capsules also incorporate propulsion motors that, when activated, allow a degree of translational

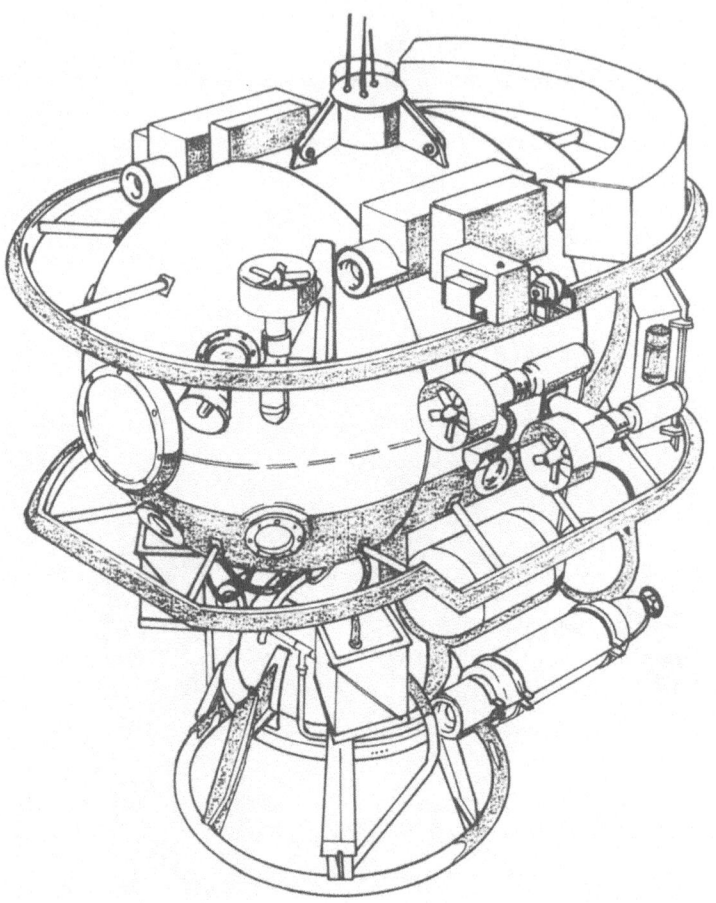

Figure 8.2 Atmospheric pressure transfer module

movement at the end of the tether to assist mating to structures
and in performing subsea maintenance tasks.

The use of such capsules for the support of the complex would
involve many of the launch and retrieval difficulties experienced
with submersibles. Although continuously connected to the
support vessel, pendulum effects on the capsule by surface vessel
movement can be a problem. The support vessel would need to be
on station directly above the complex, and maintained in position
dynamically, as anchorage systems could pose dangers to the
complex. A moonpool or cursor launch facility incorporated in a

stable platform such as a semi-submersible will probably be required.

The depths of operation may compound the problems of bell operation due to the long umbilical and control cable lengths and the displacement effect on these submerged systems by hydrodynamic drag forces. Bells may, however, have a limited application for the transport of large items of equipment that cannot be carried by submersibles.

8.3 SUPPORT SYSTEM CONCEPTS

Various support system concepts can be considered for the complex, which could provide a surface base for operations and a means of launch and recovery of the underwater vehicle. The use of a conventional submersible support vessel has already been analysed and other options are now considered (Fig. 8.3).

8.3.1 Permanently Moored Tanker (Fig. 8.3b)

A tanker may be moored permanently on the surface in the vicinity of the complex to provide buffer storage of field products, prior to trans-shipment by a shuttle tanker. The vessel may also serve as a helicopter platform for movement of personnel and materials and provide hotel facilities for personnel in transit. The tanker could also be fitted with a moonpool facility for submersible/bell launch and recovery operations and provide servicing facilities for the underwater vehicles.

8.3.2 Semi-submersible Utilities Platform

A semi-submersible platform may be permanently moored over the complex to provide services. A catenary anchorage system would probably be required to maintain the platform in position and is not recommended as the anchorage system would pose dangers to submersible operations and to the complex if the anchors were to drag.

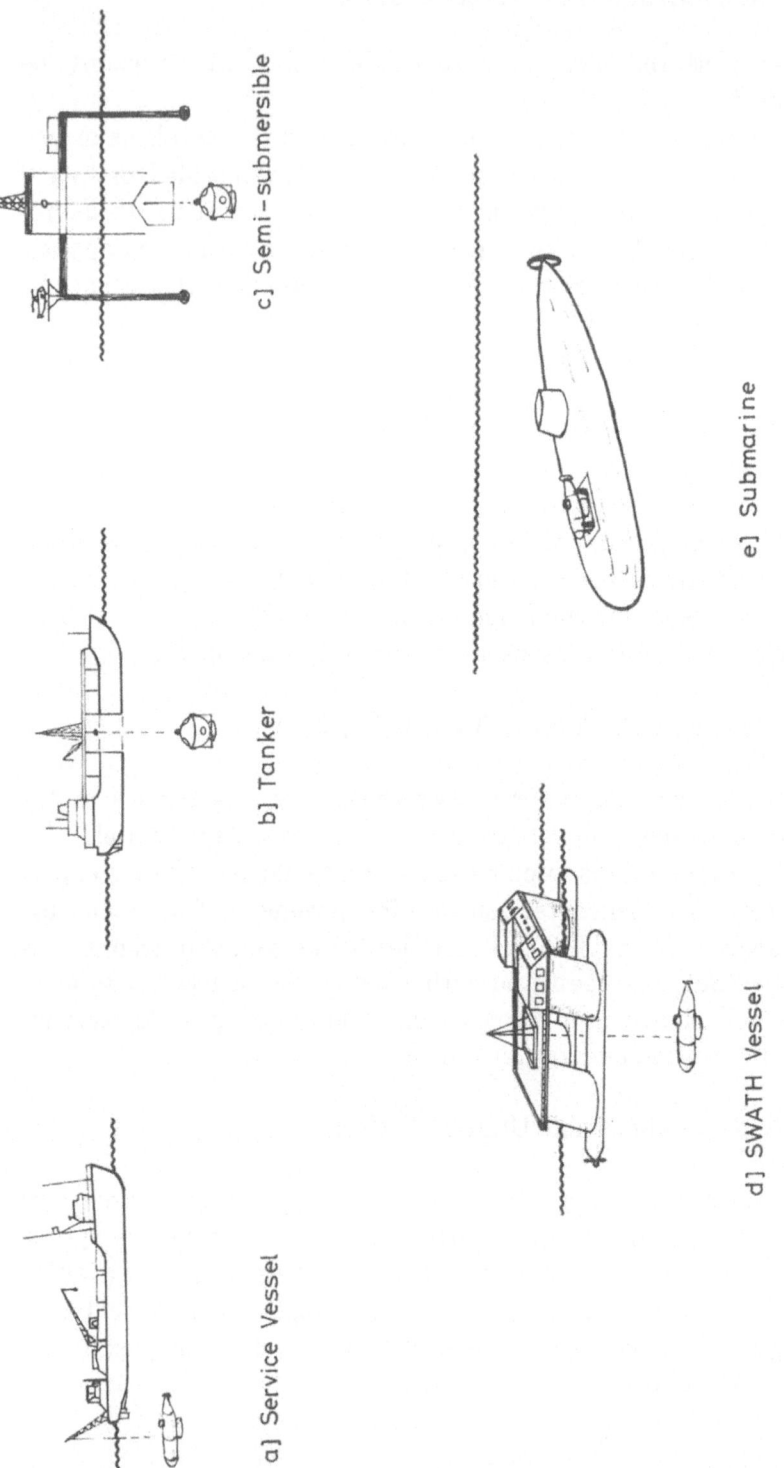

a] Service Vessel

b] Tanker

c] Semi-submersible

d] SWATH Vessel

e] Submarine

Figure 8.3 Surface support systems

8.3.3 Semi-submersible Workover Platform (Fig. 8.3c)

A semi-submersible platform will probably be required at the field
location to perform workover operations on the subsea wells and
this will probably incorporate dynamic position-keeping systems
to locate the rig on station. The platform could also be fitted with
a moonpool facility for launch and recovery of underwater vehicles
and provide surface hotel facilities.

8.3.4 SWATH Vessel (Fig. 8.3d)

The small waterplace area twin-hull ship (SWATH) is a new
concept of ship that has been developed to produce a vessel that,
at rest or underway, exhibits small motion and acceleration in
large waves. The greater portion of the ship's buoyancy volume is
below the sea surface in the form of twin torpedo-like lower hulls,
supporting the above surface structure by thin struts which are
little affected by wave action. The small waterplane area produces
only a fraction of the buoyancy change of a conventional monohull, resulting in less significant motion when in transit. The draft
of the SWATH is also deeper than a conventional monohull and
the flow through the propellors is more uniform. This gives more
efficient propellor operation.

The SWATH vessel SSP Kaimalino designed by the Naval
Ocean Systems Centre, Hawaii Laboratory, permits normal operations in 3 m waves and has an operating range of 450 miles at a
cruise speed of 16 knots. Plans are now being laid to enlarge the
vessel from the present 225 tons to 610 tons in order to provide
sufficient payload capacity, endurance and accommodation for
open-ocean work.[25]

The stability of such a vessel is very good both underway and at
rest in sea states up to 7. Such a vessel could be used as a high
speed shuttle vessel from shore to the complex location if no
permanent stand-by facility is required. It could also provide a
stable platform for launch and recovery of underwater vehicles
and, if fitted with dynamic positioning, may be able to perform
workover operations on subsea wells.

8.3.5 Autonomous Submarine (Fig. 8.3e)

The use of an autonomous submarine to support the complex from a shore base by submerged transit is feasible and negates the restrictions of surface environmental conditions. The vehicle would, however, need to be able to dock with the structure, as the size of submersible that it may be able to carry in the mothership role, may be as easily launched from the surface. The carriage of a larger submersible may be possible if more massive submarines are considered for use in the role of a power unit or as a tanker.

The development of a submarine that could dock with the structure may be excessively expensive for the operating depths envisaged. The submarine would also only provide the logistic support function for the complex and other facilities may additionally be required for export of products and workover operations.

8.3.6 Summary

It would appear that, providing the submersible size can be restrained to the present specification, there appears to be no operational restrictions on using present submersible support vessels for the logistic support requirements. Better systems may evolve because of the operational field requirements for large tankers and semi-submersible work platforms. These may provide more stable support vessels and more sophisticated handling systems for the launch and recovery of the underwater vehicles. Submersible operations would seem preferable to bell operations, because of handling difficulties of bells and the multi-function role required of the underwater vehicle. In view of the total field operational needs and transit requirements to a shore base, the SWATH vessel would appear more cost effective than the use of an autonomous submarine.

8.4 INVERTED SUPPORT SYSTEM CONCEPT

In the context of a major subsea installation such as a permanently emplaced manned underwater structure, perhaps we should not be looking purely at surface technology for solutions to our logistic support requirement. Such a new concept in ocean systems may justify a new approach to the problem.

The manned underwater complex will be relatively self-sufficient as an ocean system and will exist on the seafloor in a stable environment. We could, therefore, consider inverting our ideas on logistic support by considering the complex at the seabed as the base for operations.

In such a concept, within limitations, the underwater structure could choose when to communicate with the surface for its requirements. Transport vehicles could be deployed from the complex to rendezvous with surface vessels when conditions at the surface were suitable. Submersibles would need to lock-in to the complex for unloading and servicing, and lock-out for movement to the surface. The deployment of one-atmosphere transfer capsules from the complex would have many advantages; the anchor point would be fixed and positive buoyancy could be used with suitable guidelines to raise the capsule to the surface for recovery. A downhaul cable, driven at the complex, could recover the capsule to the structure with the resupply loads. Surface vessels, such as the SWATH vessel, would obviously need to be in attendance at the surface for recovery operations of bells or submersibles.

The application of suitable design methods should allow practical implementation of such concepts. The use of a transfer capsule for movement of personnel and materials may be possible, but may not be desirable in the overall safety design. Materials transport in capsules by such a method would be advantageous, submersible operations then only being required for personnel transport.

8.5 CONCLUSION

The use of conventional submersible support vessels should be suitable for our resupply requirements if submersible size and weight can be minimized. Submersible size could be reduced by suitable logistic planning schedules or the use of one-atmosphere transfer capsules based on the complex for materials movement. If environmental conditions are more severe than those presently investigated, consideration should be given to the use of field operating systems (tankers, semi-submersibles) for the provision of more stable surface platforms and more sophisticated handling systems. The developments in 'SWATH' vessels may provide a multi-functional role capability for our requirements, which may prove more cost effective in the context of the overall system configuration.

9

CONCLUSIONS

1. The logistic support of a manned underwater complex will require the assimilation of a three-dimensional concept of the ocean environment. Oceanic parameters will impose environmental constraints on support system operations at the surface/air interface, through the water column and at the seabed/water interface.
2. The major areas of operation will be based on the continental slope, i.e. from the continental shelf break at 200 m down to 2000 m. This represents approximately 8.5% of the world's ocean depths. The slopes, although within reasonable distance of shore, have generally been previously neglected as sites of marine installations as they are frequently in areas of high current and heavy weather.
3. Oceanographic and environmental parameters at the site location will require detailed assessment prior to the installation of the structure and implementation of support operations. Sea water characteristics such as salinity, temperature, density and currents and the bearing characteristics of the bottom sediments will need to be measured *in situ*. The local topography at the site location and the probability of volcanic or seismic activity will need to be evaluated.
4. The successful attainment of operational task requirements for logistic support of the structure will be constrained by environ-

mental conditions, support system operational limitations, production system requirements and human factor considerations. Critical operational areas will be encountered during deployment of the structure, docking and mating of underwater vehicles, ingress/egress system operation, maintenance and repair of subsea operating systems and the use of surface support systems for launch and recovery of underwater vehicles.

5. The construction and movement of such large structures will be restricted by the dimensions of the complex, the draught and hydrodynamic considerations. Suitable construction sites and deployment possibilities will be limited. Construction will need to be undertaken in the ocean of operations and limited deepwater port facilities will probably involve long ocean tows to the site location. Surface sea conditions will affect towing speeds, hydrodynamic loading on the structure during tow and surface operations during emplacement. The deployment of the structure from the surface through the water column to the seabed will be the most critical phase of operations.

6. The transport of logistic loads from a surface or subsea base to the underwater structure will be a relatively expensive operation in relation to the life time cost of the structure itself, due to the high cost investment in underwater vehicles and the systems to support their operations. The only ways of alleviating these costs are by minimizing the logistic support loads, maximizing the utilization of the logistic support systems or the use of alternative lower cost support methods. Logistic support loads can be significantly reduced by adopting liberal design features in the areas of life support, power generation, volume allocation and waste treatment.

7. On the basis of a liberal design philosophy for the structure, it is considered that an underwater vehicle of dry payload capacity between 750 and 1000 kg and a transfer hull volume of approximately 4 cubic metres is required, resupplying the structure on a multiple dive cycle of three dives a week. The utilization of the support system should be frequent enough to justify its permanent availability, while the day to day continuity of operations should not be critical to complex operation.

8. An underwater vehicle will be required that can transit from

a surface or subsea base to the underwater complex carrying the logistic payloads envisaged, perform the rescue of personnel from the structure and undertake maintenance and repair tasks on subsea operating equipment. A vehicle with an operational capability less sophisticated than the Deep Sea Rescue Vehicle and slightly better than the L5 or PC1800 submersible would appear appropriate. A twin pressure hull arrangement is recommended for safety considerations and to allow efficient use of transfer hull volume. The provision of such a purpose designed vehicle for the task requirements is considered feasible with present technology.

9. The use of one-atmosphere transfer bells deployed from the surface is not recommended for logistic support because of deployment difficulties and their limited manoeuvrability for task requirements. Autonomous submarines could be considered if they dock directly to the structure, although the high cost investment penalty may again be restrictive when compared to alternative support concepts.

10. Access to the underwater structure from the underwater support vehicle should be designed so that the integrity of the structure is retained during transfer operations of crew and materials through the interface. It is considered that an access chamber system should be used to provide a buffer area between the structure and the support vehicle, such that lock-in/lock-out transfer operations can be performed. Consideration should also be given to the incorporation of the safe haven facility in the access chamber and the use of the access chamber as a one-atmosphere buoyant ascent capsule.

11. The use of conventional A-Frame underwater vehicle support vessels should be suitable for our resupply requirement if the submersible size and weight can be restricted to the present specification. Should handling difficulties arise, logistic loads could be further reduced by modifications in resupply schedules or the use of one-atmosphere materials transfer capsules deployed from the underwater structure. Production field operating systems (tankers, semi-submersibles) may also allow the provision of more stable platforms and the use of more sophisticated handling systems for vehicle launch and recovery.

12. The 'SWATH' vessel may, in the future, provide a multiple role support vessel for the logistic support of the underwater

structure. The vessel would act as a high speed transit vessel from base to site location and provide a stable platform for hotel facilities and underwater vehicle launch and recovery operations at the site location. The vessel could possibly also be rigged to perform workover operations on subsea wells.

10

RECOMMENDATIONS

1. The design of the underwater production complex should be optimized to allow the logistic resupply requirement to be minimized. The logistic loads should be evaluated in detail once the final design concept has been determined. Resupply schedules should then be constructed in relation to the underwater vehicle's payload capabilities, limitations on surface support systems operation and available storage volume within the complex.
2. A detailed design study of a submersible vehicle to provide the payload capacity and volume allocation for the transport of the logistic loads should be undertaken. The various ancillary task requirements of the vehicle, such as rescue, maintenance and repair activities subsea, should be incorporated into the design concept.
3. A tool and instrumentation package will need to be devised for use with the submersible vehicle to measure environmental parameters *in situ* at the site location, i.e. bearing strengths of seafloor, currents and temperature, density and salinity of seawater.
4. Investigate the use of one-atmosphere buoyant ascent materials capsules that can be deployed from the underwater complex to the surface. Consideration should be given to guideline and hauldown systems, mating techniques and pick-up at the surface.

5. Perform a detailed design study of an ingress/egress system, incorporating an access chamber for transfer operations between the underwater vehicle and the structure. The study should also consider the incorporation of a safe haven facility within the access chamber and the use of the chamber as a one-atmosphere buoyant escape capsule.
6. The use of a 'SWATH' type vessel as a shuttle vehicle between the base port and the site location should be evaluated and consideration given to its use on station as a stable platform for hotel facilities, launch and recovery of underwater vehicles and as a workover rig.

REFERENCES

1. *Deepwater Oil Production and Manned Underwater Structures* by M.E.W. Jones, published by Graham & Trotman, 1981.
2. *Handbook of Ocean and Underwater Engineering*, by J.J. Myres, C.E. Holm and R.F. McAllister, North American Rockwell Corporation, McGraw-Hill 1969.
3. *Man Beneath the Sea* by W. Penzias and M.W. Goodman, J. Wiley and Sons, 1973.
4. *Conceptional Study of a Manned Underwater Station*, General Dynamics Corporation for Naval Facilities Engineering Command, April 1967.
5. *Concept for a Manned Underwater Station*, Southwest Research Institute for Naval Facilities Engineering Command, February 1967.
6. *Concept Design for a Manned Underwater Station*, Westinghouse Electric Corporation for Naval Facilities Engineering Command, March 1967.
7. *Study of One Atmosphere Manned Underwater Structures*, Ocean Systems, North American Rockwell Corporation for Naval Facilities Engineering Command, June 1968, Vols I and II.
8. *Forecast of Human Factors, Technology Issues and Requirements for Advanced Aero-Hydro Space Systems* by H. Price and J. Parker, Biotechnology Incorporated, March 1971.

9. *Safety Characteristics of Lockheed's Subsea Production System* by G. Fahlman, Lockheed Petroleum Services OTC, 1974.
10. *Safety and Operational Guidelines for Undersea Vehicles*, Vols II and III, Marine Technology Society 1974/1979.
11. *Engineering Aspects of Manned and Remotely Controlled Vehicles* by R.F. Busby, Royal Society, London, 1977.
12. *Manned Submersibles* by R.F. Busby for the office of the Oceanographer of the Navy, January 1975.
13. *Personnel and Training Requirements for the Deep Sea Rescue Vehicle* by J.F. Noble, U.S. Navy Personnel Research and Development Laboratory, October 1970.
14. *Deep Submergence Systems Terminology and Usage*, Northrop Corporation for U.S. Navy D.S.S.P., June 1980.
15. *Reliability Prediction for Deep Sea Submergence Rescue Vehicle Second Reliability Model*, U.S. Navy Marine Engineering Laboratory, March 1966.
16. *A Technical Guide to the D.S.R.V. Alvin for Use in Planning Scientific Missions*, Woods Hole Institute, December 1967.
17. *Undersea Work Systems* by Howard R. Talkington, U.S. Naval Ocean Systems Centre, published by Marcel Dekker Inc., 1981.
18. *A Functional Design of an Ingress/Egress for an Ocean Bottom Station* by Westinghouse Electric Corporation for the U.S. Naval Civil Engineering Laboratory, October 1968.
19. *The Use of Mock-Ups in the Design of the D.S.R.V.* by J. Jenousek, Lockheed Missile — Space Company, February 1970.
20. *Effective Support Systems for Transferring Personnel /Materials between Sea Surface and Seabed Production Modules* by Professor Kuo, Strathclyde University for Sir Robert McAlpine & Sons, February 1979.
21. *A Controlled Handling Method for Effective Offshore Support Operations* by Professor Kuo, University of Strathclyde, OTC, 1978.
22. *Too Complex for Contractors to Fund Alone*, Offshore Services and Technology, November 1980.
23. *Concept Study for an Autonomous Ferry Submarine for the Deep Sea Production System* by Slingsby Engineering for Sir Robert McAlpine and Sons, 1981.

24. *Access by a Free-Swimming Submersible to a Seabed Satellite Manifold Chamber* by British Oceanics Ltd, for Sir Robert McAlpine & Sons, February 1981.
25. *Ocean Technology — A Digest*, Technical Document 149, Naval Ocean Systems Center, San Diego, February 1981.
26. *A Survey of Launch Recovery Concepts and Systems for Deepstar Vehicles* by D. Usry, Westinghouse Electric Corporation, July 1967.
27. *Surface Handling of Diving Bells and Submersibles in Rough Seas* by T. Mellam, Det Norske Veritas, OTC, 1979.
28. *Stable Platforms for the Launch & Recovery of Submersibles* by N.B. Estabrock and H.M. Horn, OTC, 1972.